PIERCE *the*
DESIGN FOG

DEVELOP HIGH-QUALITY PRODUCTS FASTER
through TEAM INNOVATION

Dianna Deeney

www.pingaugepublishing.com

ISBN: 979-8-9893259-1-7 (ebook)
ISBN: 979-8-9893259-0-0 (paperback)
ISBN: 979-8-9893259-3-1 (hardcover)
ISBN: 979-8-9893259-2-4 (audiobook)

Printed in: Kimberton, PA, U.S.A.

Library of Congress Control Number: 2025914027

Ordering Information:
Special discounts are available on quantity purchases by corporations, associations, and others. For details, contact orders@pingaugepublishing.com

Publisher's Cataloging-in-Publication Data
Names: Deeney, Dianna, 1976- .
Title: Pierce the design fog : develop high-quality products faster through team
    innovation / Dianna Deeney.
Description: Kimberton, PA : Pin Gauge Publishing, 2025. | Includes 64 b&w
    photos, diagrams, and charts. | Includes bibliographic references and index. |
    Summary: Provides a playbook for concept development with cross-functional
    teams, and introduces the Concept Space Model and ADEPT Team
    Framework to gather and prioritize design inputs. The work extends these methods to translate early
    knowledge into concrete design specifications, informing later stages like risk analysis and usability.
Identifiers: LCCN 2025914027 | ISBN 9798989325931 (hardcover) | ISBN
    9798989325900 (pbk.) | ISBN 9798989325917 (ebook) | ISBN
    9798989325924 (audiobook)
Subjects: LCSH: New products – Management. | Product design. | User-centered
    system design. | Cross-functional teams. | BISAC: TECHNOLOGY &
    ENGINEERING / Industrial Design / Product. | TECHNOLOGY &
    ENGINEERING / Systems Engineering. | TECHNOLOGY &
    ENGINEERING / Technical & Manufacturing Industries & Trades
Classification: LCC TS171.D44 2025 | DDC 658.5 D--dc23
LC record available at https://lccn.loc.gov/2025914027

Contents

Part Three

This book is dedicated to all the ingenious and inventive people who take abstract concepts and translate them into beneficial products that improve our lives.

Preface

In "Portrait of Chess Players," Marcel Duchamp, the French painter and sculptor, captures two people contemplating chess strategies. They likely hold different perspectives of the same problem and create outcomes that can shift the other player's ideas, ultimately changing the course of the game.

Concept development with a team can be quite similar. It often involves multiple people tackling challenges from diverse viewpoints, each with their own goals and potential trade-offs. Their ideas can significantly influence the project's direction.

Marcel Duchamp. (1911). "Portrait of Chess Players" [photograph].

The difference is that the goal with development is to win *together*. A mutual victory means you create a product that is useful, safe, effective, and desirable for others to use.

There are two major difficulties with concept development: knowing what to talk about and managing teamwork.

It's difficult to develop something from nothing with a cross-functional team, which is one reason we jump into the solution space too early. If we create something, we can at least talk about it.

However, jumping into solutions too soon robs the team of opportunities to learn more about the potential product and the user. The more we learn before we engineer solutions, the better we'll be able to create the best possible products.

We don't need to create solutions just yet. Instead, we share knowledge with a cross-functional team and use an idea-capturing process and a systems approach that is focused on the user.

This way, it's easier to develop concepts and gather design inputs with others before we engineer stuff.

Introduction

EMPOWER TEAMS TO DEVELOP INNOVATIVE PRODUCTS

Anyone can use their creativity to design products, and we should! To manufacture high-quality products, we need a skilled team.

Product designers and design engineers in the industry connect ideas to finished products. They combine artistic vision with technical expertise. They figure out how to make products. Their goal is products that are appealing, user-friendly, efficient, and safe.

Designers don't work alone. Successful new product development requires a design concept that is developed early by a cross-functional team. A cross-functional team includes members from different functional groups within the company. These teams usually involve engineering, design, marketing, sales, and manufacturing.

Teams typically do not communicate well during concept development.

In his book *Winning at New Products*, Robert G. Cooper writes, "High-quality and effective cross-functional teams are at the heart of any well-executed project." It's a key factor that underlies success. In terms of its impact on new product performance, it correlates to profitability and timeliness, in both how fast the project was done and if it launched on time.[1]

Cooper also cites studies that show projects are more than three times as likely to succeed with "sharp, fact-based product definition before development begins." Projects have 2.5 times the success rate and earn double the market share if the team does "solid up-front homework" doing front-end activities well.[2] These activities include the fuzzy front-end and precede the development phase. This is when the market and product are defined.

However, many teams do not undertake these types of pre-development activities or perform them to the level they should. Cooper shows that only 16% of person-days on a project is dedicated to early work.[3] He writes, "In too many projects, we observed a new product idea that moved directly into development with very little in the way of up-front homework to define the product and justify the project – a 'ready, fire, aim' approach."[4]

Teams move ahead into product development without customer requirements or customer input. Comparing successful products to failed products, successful ones have about 75% more time dedicated to these pre-development activities.[5]

Little development or knowledge sharing occurs between people on a project, especially in these early phases. This results in information getting lost in translation or simply getting lost, leading to challenges in meeting the customer's wants or needs.

---

[1] "Figure 3.2," in Robert G. Cooper, *Winning at New Products: Accelerating the Process from Idea to Launch*, 3rd ed. (Basic Books, 2001), 61.
[2] "Figure 3.3," in Cooper, *Winning at New Products*, 62–63.
[3] "Figure 2.8," in Cooper, *Winning at New Products*, 45.
[4] Cooper, *Winning at New Products*, 63.
[5] Cooper, *Winning at New Products*, 67.

Misunderstandings happen.

Teams lack clear communication with designers, sometimes assuming shared knowledge. With good intentions, functional groups swap information about a new idea or opportunity they feel is complete. Even if they are done well and are thorough reports, a hand-off introduces the possibility of misinterpretation.

Others cannot easily translate the information into design. Or there are several assumptions that need to be made to jump from idea to design, and assumptions can be wrong. The information from each group is disjointed and may conflict. The designers don't know what clarifying questions to ask the team.

Designers just accept the information and move on to designing a solution without first developing concepts with their team.

When I ask development team members about their challenges, I hear this frequently: "We started a project with all the market and customer research we were supposed to do. Then, we handed it over to engineers to develop. They came back with something that we didn't want, and customers don't like!"

I also talk with engineers, and I can understand their point of view, too. "We were given all this information and had kick off meetings. We understood the problem and actually found a solution! We had to make trade-off decisions and ensure manufacturing could make it. We spent a lot of time coming up with this creative solution, and they didn't even appreciate it!"

Teamwork can be difficult.

Another reason concept development is difficult is because of challenges with teamwork. Many people believe individual work is superior to teamwork in terms of speed and efficiency—and that more participants lead to wasted time and poorer decisions. As a result, the team leaves designers out of concept sessions. Or designers decide to work alone with what they know.

These decisions can get teams stuck in a loop.

Stuck in a Loop

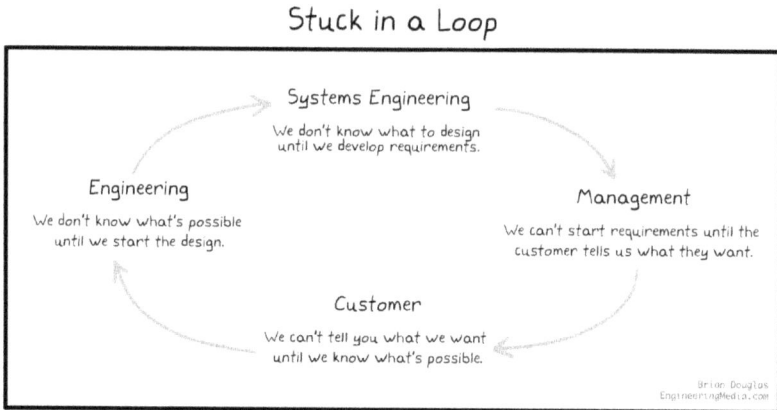

Systems Engineering
We don't know what to design
until we develop requirements.

Engineering
We don't know what's possible
until we start the design.

Management
We can't start requirements until the
customer tells us what they want.

Customer
We can't tell you what we want
until we know what's possible.

Brian Douglas
EngineeringMedia.com

"Step aside engineering, we've got this." - The Sales Department

Figure I.2. "Stuck in a Loop" by Brian Douglas

Despite the challenges, cross-functional collaboration has its advantages. We're pulling together people with diverse mindsets from across the organization to work together on a single idea. While we have different perspectives on the same problem, those perspectives can help us create a better product.

Leadership can help with challenges, like defining goals and aligning performance metrics to those goals. If leadership doesn't provide that guidance, teams can help themselves by establishing shared objectives: understanding users, their environment, the product's purpose, and company objectives.

It's not enough to just share the information. Teams need to explore it together and move ideas toward design inputs they prioritize. Designers must co-work with their cross-functional team before they create solutions. A key aspect of this teamwork is holding space for their team to share their knowledge.

Spending time on concept development matters.

Concept development is a valuable exercise. It helps us understand and link customer values to design decisions. This affects the success of our projects.

In Cooper's studies, having a "well-defined product prior to the development phase" led to projects that are 3.3 times as likely to be successful, have higher market shares by 38 share points on average, and are rated better by customers. This leads to improved profitability. Product definitions include understanding the target market, the customer (their needs/wants/preferences), and the product concept (what it will be and do, and its specifications and requirements).[6]

Better concept development also improves design inputs, which boosts design quality. Higher perceived value means higher customer spending and business profitability. Also, improved design quality leads to better quality of conformance. This lowers manufacturing and source costs and results in higher profitability.

Not all ideas are the same in the eyes of our customers, so not all design inputs should be equal. Gathering information early helps prepare for design decisions that align with your customers. The goal is not to collect all information but to focus on what impacts the customer through design decisions and related offerings.

This book is a playbook based on what works, and here's what you'll learn:

This book introduces the **Concept Space Model** and the **ADEPT Team Framework**. Used together, they are your playbook for concept development with your team. These methods enable teams to work in a way that allows them to understand both the user's needs and the company's

---

[6] Cooper, *Winning at New Products*, 61.

goals. This collaborative approach leads to the development of innovative and effective products.

You'll learn how to:

- Spend more time understanding the problem before rushing to solutions, avoiding costly mistakes later in the design process.

- Utilize the Concept Space Model to target user benefits, prevent negative experiences, and optimize the use process, keeping your user at the center of design decisions.

- Master the ADEPT Team Framework, a step-by-step approach for facilitating engaging discovery meetings that promote team involvement and consensus.

- Move beyond open brainstorming with practical techniques for collecting information and driving projects forward towards design.

- Prioritize ideas that are truly meaningful, informing not just concept development but also later stages of product development.

- Employ a systematic approach to question, investigate, and transform customer needs into targeted experiences and engineering design inputs.

This book's methods are backed by research and real-world success stories from engineering, design, project management, risk management, and quality professionals. My approaches are influenced by my own experiences as a senior engineer and certified quality professional. My career path has taken me from manufacturing operations to a specialization in quality and reliability engineering. This has given me direct insight into challenges of product development, like communication breakdowns and translating user needs into effective engineering inputs.

Collaborating with a team can be very rewarding. Launching a new product together builds camaraderie. By challenging your own ideas, you'll get better ones. You might find other business aspects interesting and develop a new passion. And a shared responsibility toward success leads to bigger celebrations and a desire to do it again!

Being able to see your product ideas used by happy customers is very exciting and satisfying! You positively impact others' lives. You bring joy, make tasks easier, or help people work better. These impacts spread to others, like ripples in a pond.

By the end of this book, you'll have the right mindset, tools, and strategies to better work with others and translate their knowledge into design inputs. This book helps you move that product from idea to reality.

Visit the website *piercethedesignfog.com* for additional templates, models, and other products that complement this book. You will also find my contact information, so you can reach out if you have questions.

Why this book applies to you:

This book combines best practices, research, and practical experience to bridge the gap between product idea and engineering design. It uniquely combines methods for running effective discovery meetings and developing design concepts and design inputs.

The methods in this book apply to new product development.

Products may be tangible or intangible and may involve services. Customers can be external, such as business customers who purchase our products in the market. Customers may also be internal, like a team tasked with designing new in-house software. If you have a product idea that addresses a customer's needs, then you can apply the methods in this book.

This book helps people who are designing.

Managers and leaders: If you're managing designers through product development, then this book helps you identify and plan activities that boost your team's concept development.

Designers: If you're a designer, this book provides you with a plan and a method to co-work with your cross-functional team. Some professionals distinguish between a product designer and a design engineer. A product designer may identify more closely with work that is focused on user experience and human-centered design, while a design engineer works more closely with material selection and design for manufacturing. Both designers benefit from concept development.

Cross-functional teammates: If you work on a team for product design and struggle with communicating what you know, then this book will give you a plan and a method to work with the team, including designers. You'll not only share your knowledge but also learn how to lead the team forward in the product development process.

The scope of this book lies between user needs and design requirements.

This book applies after the team has identified at least some primary user needs. User needs, or customer needs, are defined and described in this book to differentiate them from targeted user benefits. The methods outlined in this book can help the team in identifying more needs. This is because the team is gaining a better understanding of the product opportunity by collaborating.

In this book, we focus on developing design inputs, which can be documented as a list of written requirements. However, this book does not provide guidance on how to word these requirements. Teams should determine the best documentation methods for their work and results, ensuring future reference and design control.

The methods in this book complement your existing systems and work with a variety of approaches.

Many designers work within a product development process. This helps guide teams, controls the process for review, ensures the project is meeting targets, and coordinates departments. Your company probably also uses a quality management system, which helps ensure people perform duties to ensure quality, quality measures are created and monitored, and risk management is part of decisions.

Should your systems already cover concept development, this book provides additional support. Even if your systems don't include it, this book's methods will still work with your design process. The methods in this book may help you become more effective and focused on your current work or new activities. In either case, you shouldn't need to work outside of your existing systems to apply these methods.

This book offers many activities—try them all at once or one at a time.

You don't have to tackle all the activities simultaneously to enhance teamwork and design. You can improve your team's concept development by strategically choosing activities. Instead of feeling overwhelmed by trying to do it all, you can make progress by addressing specific challenges one by one.

For examples, you can choose to first adopt the ADEPT Team Framework for any meeting where decisions are needed. Or you may want to do activities related to symptoms as input into your risk management efforts. If you're far along in your current project, you may want to first do one of the activities in part 3, where we're developing design inputs.

This book lays out a system that works together but it's not all or nothing. Just do what you can, wherever you are. In that way, you embrace the quality mindset of continuous improvement.

What this book does not do:

(See the Appendix for additional resources.)

This book will not help leaders decide whether to pursue a new product project.

This book does not include information on how to identify new market opportunities. Tasks such as conducting market research, technical feasibility studies, financial analysis, and other related analyses are not covered within the scope of this book. Some of the information in these analyses may influence concept development, however, so team members should review these analyses.

This book does not teach all facilitation skills.

The techniques in this book will improve your skills as a facilitator in concept development. Factors that contribute to success include understanding product development, having a cooperative mindset, planning collaboration, and setting goals for team meetings. If you're reading this book, you probably already know about product development. This book focuses on mindset, planning, and goal setting for effective collaboration.

While this book provides valuable insights and guidance, it does not cover every skill you may need to facilitate teamwork. This book does not cover coaching, conflict resolution, interpersonal skills in a group environment, or behavior management. However, common facilitation skills intersect with these areas.

These skills include active listening, effective communication, empathy, problem-solving, critical thinking, creativity, flexibility, and patience. To develop these abilities, you must recognize their value, identify your weaknesses, and actively practice and seek feedback to improve. I recommend you supplement this book with other resources that address these aspects of facilitating.

Once you have the tools to be a successful facilitator with your team, the best way to get better at facilitating is by doing it.

Cross-functional team creation and management are beyond the scope of this book.

This book does not provide guidance in assembling and supporting cross-functional teams, which are essential in today's business environment. Setting up and developing these teams can be challenging due to various dynamics.

One crucial aspect of team management is understanding group behavior. A wide range of personalities, motivations, and communication styles are represented on teams. Understanding group dynamics lets team leaders manage conflict better, communicate more effectively, and create a positive team environment.

Managing the various stages of team development is important. You may know these phases as forming, storming, norming, and performing. They describe the natural progression of a team from its initial formation to becoming a high-performing unit. By recognizing and addressing the challenges that arise in each phase, team leaders can guide their teams towards success.

Teamwork relies not only on team leaders but also on organizational support. It is important to create a respectful environment where all team members feel valued and heard. This fosters collaboration and innovation. Supportive leadership and reporting structures provide cross-functional teams with the resources and authority they need. Companies can provide training to improve employee skills and teamwork.

If you want to develop better ideas with your team before you create solutions, then this book will show you what's important and why, who should be involved, and how to do it. Our goal is better information for design.

So, let's take off our problem-solving hat and put on our problem-finding hat.

We begin with a product idea.

Part One

MASTER THE PRINCIPLES OF TEAM-BASED CONCEPT DEVELOPMENT

IN PART 1, we lay the foundation for working with a team to develop a concept for design inputs. In this part, we:

- More fully define concept development

- Identify common problems during this phase of product design

- Review the benefits of cross-functional teamwork and how to overcome its challenges

- Introduce how to use a system to model a concept for knowledge sharing before we even have a design

- Explore how models and templates are visual aids for the team

- Explain the benefits of a consistent approach

When we design, we're faced with a world of opportunity and choices. But we're not designing products for the entire world. We're designing with a specific aim: the use space and our customers. During concept development, we're trying to better understand the space and the people.

We don't want to only rely on wild guesses or our imagination, prototyping idea after idea only to have them shot down, scrapped or picked apart. We also don't want to tie down or stifle our creative process. After all, we are creative people. But when we meet with our cross-functional team and our customers, we may not be sure what to ask them or what information we need. They may not be sure what details are important. Something that seems obvious to them based on their experiences might be a revelation to someone else.

Understanding the concept space is our first step before design or ideation. It includes the use environment and customer expectations for this new product. Our team can help us prioritize decisions or compare ideas against each other. This helps us figure out which ideas are going to be better for our product design.

Concept development's main purpose is to share knowledge.

Concept development is an early stage of new product development. It happens after a business approves a project but before designers create solutions. The team generates, evaluates, and refines ideas. Goals are to identify promising product ideas, prioritize them for decisions, and develop them into more detailed concepts. The result of concept development is information for design inputs.

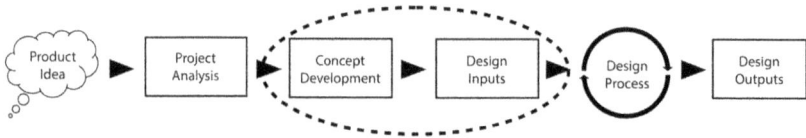

Figure P1.1. Concept Development's place in Design Development

Design inputs result from everything learned during the project. These inputs guide the team through the design process. Design inputs can include requirements like function, performance, and safety. Design inputs may also include user preferences. We link these design inputs to user needs and other constraints, ensuring the final product meets needs and expectations.

The **design process** is a cycle of design. It involves deep work, trade-off decisions, and specific solutions.

Design outputs result from the design process. Examples of design outputs include technical specifications, drawings, three-dimensional models, and prototypes.

When creating products, we consider many things about our customer.

The **use space** describes the environment in which customers will use our product. It includes when, where, how, and who uses our device. It also includes the type of interactions our product may have with other products. This includes understanding environmental sources of stress, like high temperature and vibrations.

The **concept space** is a combination of our undefined product, the use space, and targeted customer experiences with our product.

When we think of creating products, we most likely think of the part of the development process that is inputs, process, and then outputs. Yet, more work needs to be completed before we even get to the design inputs step. I view the whole product development process as a way in which we learn more and more about an idea as we develop it. The more we learn about it early in its development, the better we'll be able to make sound decisions.

We also want to facilitate some of the narrative and be involved in the concept discovery. This is because we're actually gathering information that we need later for design. Sharing knowledge gets everyone on the same page.

If we don't facilitate discovery meetings, no matter how often we meet with our teams, we won't have shared what we need to about concept ideas.

Knowledge sharing in typical meetings isn't enough, and regular check-ins don't work.

Most of our meetings have an agenda with talking points. This is acceptable when we need to review information and make decisions. But it is not enough when we're trying to gather information for a concept design.

We may think we can follow up with individuals to get concept information, assuming it's easier to meet with one person and ask questions. However, when we choose to do this instead of meeting with the team and facilitating discussions, we're going to miss potentially critical information about our concept product or service.

I was once given the lead to design packaging for medical electrical equipment accessories, including products with batteries. It involved testing to standards for shipping and handling, and I worked with various custom-cut foam and corrugated boxes. My approach was to gather design inputs from individuals, one by one. I created requirements documents as I went, and they were reviewed and approved. I felt things were going well.

Early one week, I communicated with suppliers about critical specs, finalizing them for production. By late Friday, a colleague called out to me from a nearby cubicle in the same row as mine.

He smiled, leaning back in his office chair with his hands folded across his chest.

"You're working on the packaging, right?"

"Yes."

He nodded. "I meant to tell you earlier, but I never saw you," he said. "It's important."

What he told me derailed the design decision I had made earlier that week. My jaw dropped.

"That is really important!" I said. I explained what I had done. "Why didn't you tell me about this when we talked before?"

He shrugged. "I forgot, then I didn't see you for a while. I didn't think you'd be so far ahead."

I sighed. "You were in the weekly meetings when I gave my updates. You've known about this for longer than that."

He shrugged again.

"Then why didn't you just email me?" I asked. He gave me another excuse.

I had relied on regular and individual check-ins with people. If I instead held a few working and discovery meetings with the team together to discuss the concept and design inputs, I would have significantly improved the chances that everyone received and understood this important information. I could have guided the team through various aspects of the concept space, focusing on one aspect at a time.

Setting aside a specific time to talk about this would also have helped my team focus on the product. The meetings would provide opportunities to voice ideas and share information.

We must hold space for focused conversations with our cross-functional team.

First, when we have a focused working meeting with a cross-functional team, they bring their whole selves: their diverse viewpoints and full range of experiences that relate to what the customer wants, the characteristics of the use environment, and what's best for the business.

Second, when we facilitate a working meeting, we provide scope and a common focus, which then provides an opportunity for the team to address important information that otherwise might get forgotten or overlooked.

We cannot control what other people think, say, or do. Or when they do it. What we can do is provide an opportunity for discussion and focus. When we're looking for information about a concept, providing working meetings to explore the use space provides the team and the designer that chance.

Figure P1.2. Idea communication includes feedback
photo by Freepik and vector by kreativkolors / Freepik

Relying on our design control process or product development process is not enough. We also cannot call a meeting without a plan and hope it works out. We need a plan to work with others and gain the targeted information we need for design inputs. This helps them share their knowledge and helps us close the feedback loop.

Chapter One

WHY EARLY SOLUTIONS KILL INNOVATION

Projects are often started by jumping straight to a solution, even a specific technology. That's the wrong place to begin. You want to start by asking questions and considering alternatives. At the outset, always assume that there is more to learn. Start with the most basic question of all: Why?

Good planning explores, imagines, analyzes, tests, and iterates. That takes time.

- Bent Flyvbjerg and Dan Gardner, leading experts on project management and decision-making and authors of "How Big Things Get Done"

⸺⟨⟨⟨⟩⟩⟩⸺

THIS CHAPTER introduces the idea of staying in the problem space before designing solutions. We also emphasize the importance of defining the problem before jumping to a solution. Finally, we highlight the need to understand the gap between the customer's current state and their desired state after using our product.

A new product or service offering starts with someone giving us a problem to solve. In new product development, it's an opportunity the company has identified. Someone has talked with customers, vetted an idea, and identified customer needs. Customer needs exist in the problem space. We are solving a customer problem. The solution is a product for them to use.

We collect the information and begin designing. Our design process happens alone, at our office cubicle or bench. Our product is spectacular and clever. We even worked with other engineering disciplines to make it happen.

Then, we present it to the people who discovered the customer needs and identified the opportunity. Puzzled, they say, "What is this? This isn't what we asked for. This isn't solving our customer's needs."

"What do you mean?" we respond. "It meets the needs you told us!"

We spent so much time developing this product. We're invested in it and believe in it. We spent resources producing prototypes. The manufacturing group has identified preliminary production processes. It's nearly done.

This is what I call the Ta-da Flop. We said, "Ta-da! Here's the design!" But it went "Flop" with our cross-functional team, like a deflated balloon. It wasn't what they needed.

What went wrong in this situation? It happened way back in the beginning. We jumped too soon into designing a solution to a problem. This jump was full of big assumptions. We thought our team told us everything we needed to know to design a solution.

Instead, before we create solutions, we must develop concept ideas with our team so we can be efficient and effective. Before we design things, we need to talk with our team. Even before that, we need to stop and think about how to be strategic.

Every development project has at least two development spaces: the problem space and the solution space.

One third of creating new products is creative and artistic. The **solution space** is closely associated with what we're developing. It is where we apply our designing, engineering, and innovation skills.

The other two-thirds of creating new products are about solving problems. The **problem space** is associated with our customers' needs and desires. We want to discover and then solve problems with our product.

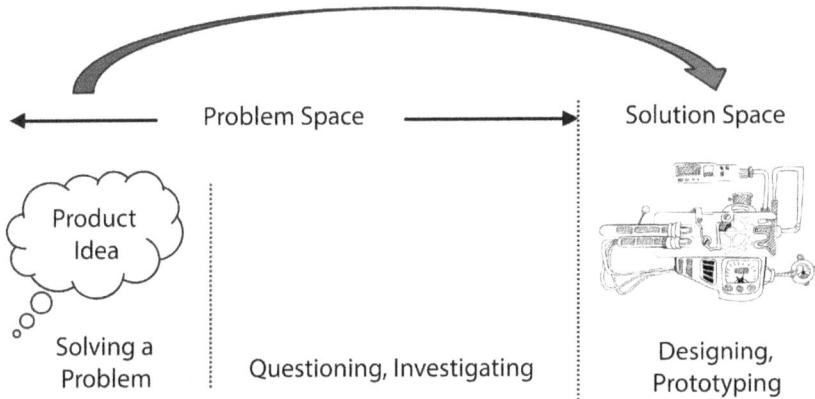

Figure 1.1. The two parts of the problem space versus the solution space industrial design doodle designed by macrovector_official / Freepik

Problems arise from the perspective of our customers:

- We create products because customers have problems. They reach for our product to solve their problem.

- Other problems involve our products' performance, which then translates back to our customers. Our product is not performing as our customers think it ought to. Or it is not safe in ways our customers demand.

- It's not always about big problems. Our customers may like our product well enough, but we want to change aspects of it.

In all cases, we're identifying a problem our product will address.

Recognize that problems are a big part of the development process. This allows us to measure design progress, make assessments of different choices, and to ensure we're customer focused. Targeting problems is practical, not pessimistic.

Teams uncover details about the problem space by doing customer-focused work. This work helps us identify potential projects and approve them for development. They do activities like:

- Identify preliminary needs

- Conduct a market needs analysis and technical assessments to determine if we can develop this type of product

- Host focused user groups to get "voice of the customer" data

The results of our work are a very rough product concept that solves a customer's problem. If we've done a good job of identifying the need for our product in the market, then we'll have gotten a good start at defining the problem space.

Through these activities, we discover what it is we're trying to accomplish with our product. We have answers to what gap our product's use is going to fill. Teams create and analyze this information to determine if we should pursue a project. But too often, we make the mistake of jumping right from these preliminary assessments into creating design inputs and designing the product.

There are two phases of a problem space: identifying the product idea and questioning/investigating.

The problem space is more than just identifying a product idea. In classical problem solving, we describe our problems in two parts. One part is the **problem statement**, which identifies what is wrong with what. The other part is the **problem description**.

A problem description includes details like timing, magnitude of the problem, location, and any evidence related to it happening. It helps provide direction for what actions we take, including what we analyze as part of our investigation.

We need both the statement and description to solve a problem. Those who solve problems know that time spent defining the problem is work we do toward gaining a solution, without actually creating solutions.

For new product development, our product idea is like identifying a problem statement. It is the gap between where our customers start and where they want to be after using our product. If our customers do not have a problem, then they're likely not going to need our product. A product idea solves a problem and can be as simple as a goal—for instance, *"We will develop a product to help bikers store their bicycles in their living areas."*

Once a business approves a project, we've already started down the path of having a product idea that solves a customer problem. That is just the first phase of really defining our product. Further problem exploration requires more questioning and investigation.

Our goal is to investigate customer experiences and question assumptions about our concept product. We want to understand information like the different users and their use environment. We want to target customer experiences and gain insight into usability and safety. These are all considerations of the concept space.

To help you visualize this, let me introduce you to the Double Diamond Design Process. This diagram was popularized by the British Design Council in 2005, originally for software developers.[7] I've modified it for our use.

Diamonds represent two cycles of work. The diamond's shape represents the breadth of ideas. Divergent thinking is when we expand ideas and

---

[7] "Eleven Lessons: Managing Design in Eleven Global Brands," Design Council, accessed March 20, 2025, https://www.designcouncil.org.uk/fileadmin/uploads/dc/Documents/ElevenLessons_Design_Council%2520%25282%2529.pdf.

explore them to find information. Convergent thinking is when we retract ideas by prioritizing and making decisions.

The dark highlighted areas in the diagram represent the questioning/investigating phase of our problem space, which is where concept development happens.

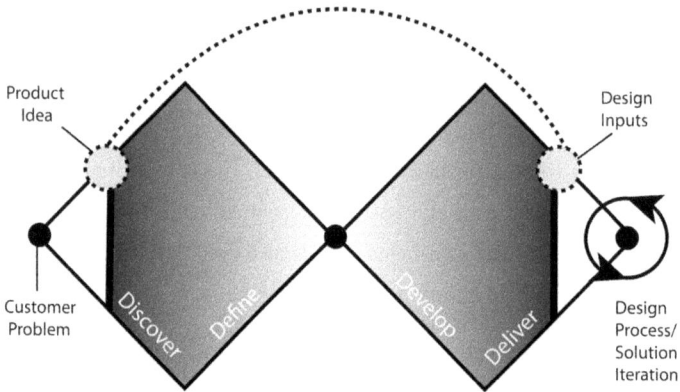

Figure 1.2. The Double Diamond Design Process, adapted

Teams frequently skip or abbreviate this process of questioning and investigating. We don't do as much teamwork as we really need. Instead, we jump right to the delivery phase to make stuff. When we jump from a product idea to design inputs, it's like jumping across the peaks of the diamonds. We skip the define and develop stages altogether.

We also cut short the discover and deliver phases, though there is a lot of worthwhile exploratory work to be done between identifying the problem and coming up with a solution. When we skip these steps, we also miss out on half of the project planning opportunities. CEOs, architects, quality professionals, and usability experts recognize this problem. Donald Reinersten, authority on product development flow, and Stefan Thomke, expert on innovation strategy, wrote in 2012 for *The Harvard Business Review*:

"Articulating the problem that developers will try to solve is the most underrated part of the innovation process. Too many companies devote far too little time to it. But this phase is important because it's where teams develop a clear understanding of what their goals are and generate hypotheses that can be tested and refined through experiments. The quality of a problem statement makes all the difference in a team's ability to focus on the few features that really matter."[8]

Skipping these steps may look like a team reverse-engineering requirements to fit customer needs. Or the project manager is drip feeding information from the early phases to the product designers. We may create barriers by the way we handle information and whom we assign to teams or tasks. We also may experience this:

- I create too many versions of my product before it's finally done.

- I have a hard time getting feedback from my team.

- My work is nearly done and then it gets picked apart by everyone.

What's most frustrating is when these products don't pass enough validation tests with users, forcing us to redo everything.

Concept development is a process for questioning and investigating that supports iteration.

New product development is not a linear process. Some models may suggest a linear progression of ideation to creation, but within those models is a lot of iteration, adjusting, and changing as we learn new information.

When we work in new product development, we're really learning about this new potential object as we develop it. We shift, adapt, and change our ideas about the product as we learn more about it. We want to

---

[8] Stefan Thomke and Donald Reinertsen, "Six Myths of Product Development," *Harvard Business Review*, May 2012, https://hbr.org/2012/05/six-myths-of-product-development.

learn as much as we can as early as possible because what we learn about our product affects our decisions.

To "design the right thing the first time" is unrealistic and leads to teams being risk averse and failing to innovate. If a team feels pressure to not change decisions, and ideas are not tested, then they are going to choose safe options that they know will work.

To encourage teams to innovate and give them space to test out new ideas can help them learn more about the product early in the design process. Through questioning and investigation in the problem space, we develop concepts. These concepts are about a product idea and how it relates to the customer. This is all prior to developing solutions.

It's easy to be misaligned on concept ideas.

Others on our team can think requirements are obvious, even if they are not. We must be thorough when examining information for a concept product. This is especially the case today with the fast pace of innovation and rapidly developing standards of design, human factors, safety, or regulations, just to name a few. Not being aligned with the expectations of decision makers can derail the product development process. The decision makers whom I refer to now are the ones who can veto a design concept.

We might be less careful about uncovering hidden design inputs when we're revising an old, successful product. Sometimes we pursue new applications for existing products. This initially seems like a good idea because we already have something that works. We just need to change our existing product.

This was the case for a hand-held device with a trigger. The existing product had many years of success for one application in the field. It had product name recognition—like the brand name Kleenex® has for tissues. Customers in the field would see the product and instantly know what it was.

The company wanted to pursue a newer field application. It was a similar use, but the purpose was different enough that it still had to be proven. Since the original design of the device, human factors had become more of a focus. Updated standards also required new analyses. The project manager hired a third-party design team to develop an engineering concept for the human interfaces of the product.

The company did extensive research to define the product idea that would solve a customer problem. They did customer research, including the use of a Kano Model for customer satisfaction. Company personnel organized focus groups, conducted individual interviews, and did a thorough market analysis. The third-party design team had access to all this information.

The design team then worked on engineering a solution. They proposed a brilliant design: a box motor with a foot pedal. It met the current standards of human factors, was easy to use and make, and met other regulations. However, they had missed a critical design input. The company wanted to continue the brand recognition with a hand-held device with a trigger. This totally upended the work the design team had done, and they needed to restart because they missed this critical design input. This design input was not implicit but assumed.

Pausing in the step between a product idea in the problem space to actually engineering a solution may have helped the team capture this hidden design input. Before they created a box motor with a foot pedal, they needed to explore the concept space more with the decision makers to ensure their expectations aligned.

Concept development bridges user needs and design inputs.

When we evaluate concepts with our team, a needs assessment has probably already been done. The business has identified an opportunity and started a project.

Needs are gaps within the problem space that people in our target market have, whether they're really aware of them. We've already identified needs. That's what started this whole product development project. Our project goal is all about developing a product our customers will use to fill the need.

User needs describe a gap.

User needs fill a gap between what is now and what the user wants in the future. Our project goal is all about developing a product our customers will use to fill a need. We're not only providing a device, but also enabling our customers to use it.

A needs statement can be worded like this:

> [Our customer] needs [to do this] so they can [achieve this].

If we are developing a bike stand, an example of a functional needs statement is:

> "[Rick, a busy biking enthusiast,] needs [to safely store his bike in his apartment] so he can [secure his bike from damage, prevent damage to walls, free up floor space in his apartment, and perform bike maintenance.]

An example of an emotional needs statement is:

> "[Rick] needs [to think the bike stand is sturdy] so he can [feel the bike and his other belongings are secure and protected]."

We can meet the needs of our customers in several ways. Needs capture what the customer, using our product, must accomplish. They don't describe how a thing is to be accomplished. We make trade-off decisions during design with the needs as our target. We validate with our users that what we designed met their needs and filled the gaps we identified.

Design inputs describe what the design must include.

Design inputs are what we use as guardrails against which to design a solution. These are all design inputs:

- Listing design requirements, including standards and regulations

- Identifying scenarios and features associated with risks, including safety and reliability

- Understanding how our users perceive, understand, and act when they use our product

We use those design inputs to design a solution and test to verify our solution meets the expectations of the design inputs. We can develop design inputs by using the information we gather in concept development.

Concept development is what we do with a team in the problem space to develop ideas.

Concept development is not a statement or goal, but a collection of ideas focused on achieving a customer's need with a targeted customer experience. We start to explore and describe how our offering may meet that need and experience. By gathering and examining ideas for features, interactions, and service options, we identify targets and explore potential options that support those targets. We prioritize options based on customer experiences. Even though we're setting targets for customer experiences, we're also analyzing design ideas.

Coming out of concept development, we have learned more about our users and the product to make decisions for design inputs. Activities from concept development can help us update our user needs and come up with design inputs.

Target additional positive user experiences.

Some user needs statements, like an emotional need, target the user experience. However, not all user needs statements reflect user experiences.

The user experiences we want to target for our product development may differ from the user needs. Instead of being user needs themselves, they support or supplement them.

For example, we want Rick to have a certain feeling about the bike stand when he assembles it. We want Rick to think the bike stand is sturdy so he feels good about leaving his bike in his apartment the first time he uses the stand. We've also decided Rick needs to assemble the bike stand so he can carry it up the stairs to his apartment and fit it through the door. With our design, we may want to target a positive user experience, like this:

"[Our customers] can [assemble the bike stand within minutes of receiving the box] so they can [use the product quickly, gain confidence in their decision, and have a sense of accomplishment.]"

While Rick needs to know the bike stand is sturdy when he assembles it, we've discovered we also want him to be able to assemble it quickly. We may not have highlighted that as a need, but it is a targeted user experience. Depending on our design control process, we could add it as a need that we've discovered during concept development, or we could develop it as a requirement.

Some positive user experiences will not fit as a need. If this user experience is a true target for your team and has been identified in concept development, then you should adopt and develop it in whatever way makes sense with your design control process.

Examine negative experiences.

We may also want to avoid negative user experiences. Negative experiences are things we try to design to avoid, remove, or limit. This is a different design target than trying to create a positive experience.

Teams can twist some negative experiences into positive ones, but there are reasons not to do that. Designing-out negative experiences will lead to different design solutions and different options because we learn more about what leads to that negative experience and can address it. In addition, a negative experience could lead to stronger emotion. Negative experiences, when they occur, may relate to risks and be part of risk assessments. We'll hear about some of them through customer complaints.

A negative experience may sound like this:

"[Our customers] may [not be able to fit their bike to the rack], which leads to [frustration]."

To test out how we think differently about positive experiences versus negative experiences, I inputted two prompts into an AI large language model:

> **Prompt 1, negative experience**: What are potential design features and service options to avoid this customer experience? [Our customers] may [not be able to fit their bike to the rack], which leads to [frustration].

> **Prompt 2, positive experience**: What are potential design features and service options to meet this customer experience? [Our customers] need [to fit their bike to the rack] so they can [feel satisfied].

(The results are shown in detail in Appendix A.)

The customers' negative experience led to solutions like:

- adjustable rack
- universal compatibility
- clear labeling and instructions
- safety features
- customer help
- services
- return policies
- feedback mechanisms

Our customers' positive experience led to solutions like:

- intuitive design

- easy-to-use mechanisms

- secure fit

- durable construction

- different service options

Not only are the design ideas different, but we're also focused on different solutions: reduce customer frustration or foster satisfaction. Reducing customer frustration is a stronger, more specific emotion that, in this case, led to more specific concept development.

Design for usability.

The use process is an important factor when designing products. Our customers actually experience our product during its use, not just after the job is done and they've achieved their goals. User interface, usability, and human factors are all design inputs.

Early standards on user-centered design focused on user interfaces with software applications. However, the idea of usability has expanded to all user interfaces.

The International Standard (ISO) 13407—"Human-centered design processes for interactive systems"—was written for computer-based interactive systems. It has since been withdrawn and replaced with the ISO 9241 series of standards, which have a broader scope.

The accepted definition of usability for User Experience (UX) workers is from ISO 9241-11:2018: "Usability: The extent to which a product can be used by specified users to achieve specified goals with effectiveness, efficiency and satisfaction in a specified context of use."[9]

A generic process for usability was defined in ISO 13407, shown in the diagram. It follows a typical design process, but with context of use added at the beginning. Here, usability is built into a typical design process or was a parallel design track. However, people found it did not adequately address different users and their needs/wants and did not help with determining measures of success.[10] Just adding usability specs to a design process was not enough.

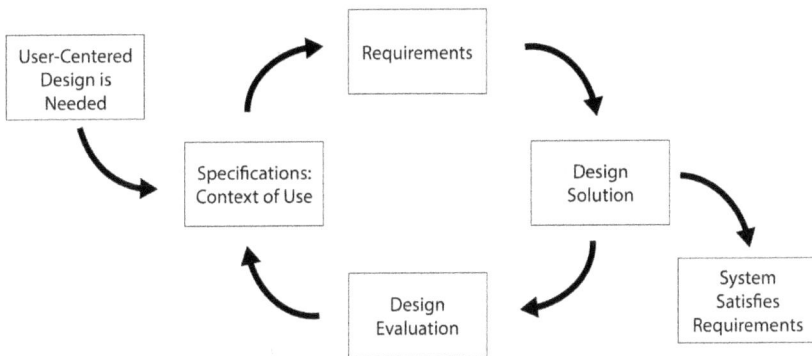

Figure 1.3. Activities of a User-Centered Design, adapted from ISO 13407

During concept development, we clarify different users and their needs/ wants and develop measures for usability. We can look more closely at the steps our users may take to use our product as a step toward usability in early product development. This information can help us develop ideas about design inputs. We can work to better understand which steps are critical to perceived quality, what adds value, and (if there are many users) who is doing what step and when.

[9]   Timo Jokela et al., "The Standard of User-Centered Design and the Standard Definition of Usability: Analyzing ISO 13407 Against ISO 9241-11," *CLIHC '03: Proceedings of the Latin American Conference on Human-Computer Interaction*, (2003): 53–60, https://doi.org/10.1145/944519.944525.

[10] Jokela et al., "Standard of User-Centered Design".

The ADEPT Team Framework and the Concept Space Model provide a solution for team collaboration.

The **ADEPT Team Framework** and the **Concept Space Model** are the solutions presented in the rest of this book. The framework is a process, and the model is a targeted tool for discussion.

Together, they make up a process to develop customer needs into targeted customer experiences and design inputs. Using these between ideas and solutions, we can stay in the problem space longer. They help us avoid jumping into creating solutions too early without focusing on the customer.

The ADEPT Concept Framework with the Concept Space Model helps teams collect, test, and iterate ideas together. We invite our cross-functional team to explore the ideas with us. Customer insights can fuel ideas about features and offerings. We prioritize these ideas based on targeted customer experiences.

The model is a targeted tool that helps teams identify what additional testing should be done. Early testing can be as simple as interviewing customers or performing research. The framework and model help our team develop a more complete understanding of the problem and generate ideas. This is how we learn about the use space and its users. From this, we can develop design inputs and create the best solutions possible.

Key Takeaways:

1. It's essential to define the problem thoroughly before seeking solutions.

2. Concept development should focus on understanding the use space and the customers.

3. Avoid jumping into solutions too early.

Reflection Questions:

1. How often do I find myself or my team jumping to solutions before fully understanding the problem?

   _____

   _____

2. What are the consequences of not fully exploring the problem space for us?

   _____

   _____

3. What are some strategies for staying focused on the problem space? What do we do now? What have I seen others do? What are new ideas I can add?

   _____

   _____

Chapter Two

UNLOCKING TEAM SUPERPOWERS

"If your team agrees on everything, working together is pointless... Let go of the idea that all conflict is destructive and embrace the idea that productive conflict creates value."

- Liane Davey, Ph.D., New York Times bestselling author and leading authority on team effectiveness

———⌖———

IN THIS CHAPTER, we discuss the importance of co-working with a cross-functional team. We emphasize facilitating meetings to gather design inputs and develop product concepts. We conclude with how to interact with our team for communication success.

A cross-functional team includes people who are assigned to a project. They each represent a certain function of the organization. They usually have decision-making capabilities for the project and the product itself.

Cross-functional team members may represent manufacturing, marketing, reliability, and quality. Sometimes it's helpful to view project management as part of your cross-functional team. These people bring a different perspective on product development.

Working with others to develop ideas is difficult.

One challenge of working with a cross-functional team is that there's a perception that too many voices impede productivity. Many individuals feel they would be better and faster at the job on their own, rather than as part of a team.

They may feel the result is wasted time and poorer decisions, especially as the number of participants increases. "I just need to know what I need to know so that I can design something already. And then we can move on with the project," they think. Not making progress on a project is a real concern. And if we have many decision makers, that makes it more challenging.

People often try to avoid conflicts, but conflicts are unavoidable. When individuals collaborate, conflicts are bound to arise. These conflicts usually stem from differing priorities, a lack of project leadership, or a lack of a unified working system.

Within the challenges of cross-functional teamwork also lies its strengths.

We can overcome these challenges, and we need to, because there are too many benefits of cross-functional co-work we don't want to ignore. We're pulling together people with diverse perspectives across separate functions of the organization to work together on a single idea. They have unique life experiences and viewpoints and may not have much in common with each other besides the project. They represent a certain function of the business, which has its own goals. These differences are all benefits for the company and our customers.

They give us a more complete picture, especially when it concerns the use space of a new product. Each person's viewpoint helps create a more complete vision of a product's performance and the risks it may introduce to the user.

"Involving multiple roles, for example users and developers in the risk identification process, will result in a more complete set of identified risks than if only one role is included in the process."[11]

Cross-functional collaboration is powerful and provides clarity. Interacting with others helps you understand goals and project objectives. Good collaboration leads to better alignment and improved designs for teams.

Cross-functional collaboration has another benefit. As a product designer, you have many internal customers. Your responsibilities include decisions on appearance, manufacturing, and materials. However, it takes a team to turn a design idea into a tangible product. The designers' choices influence many others, especially those on the cross-functional team.

Collaborate with the team to gather information and develop concept ideas.

To uncover ideas, it is important to seek information actively from the cross-functional team. Not working with the cross-functional team and failing to ask questions is a missed opportunity for designers. In the solution space, daily decisions are made that impact all aspects of the product, such as manufacturing, distribution, users, and disposal. By reviewing design inputs and taking part in discussions, designers can make more informed design decisions.

I have worked in situations where experts created user information and stored it in a folder for designers. The designers didn't understand it, made their own assumptions about details, or didn't look at it. This process was challenging, and we overlooked important design features that could have benefited our customers. I also spoke with a usability experience expert who shared his experiences working with designers. He mentioned

---

[11] Christin Lindholm and Martin Host, "Risk Identification by Physicians and Developers: Differences Investigated in a Controlled Experiment," *2009 ICSE Workshop on Software Engineering in Health Care* (2009): 53–61, https://doi.org/10.1109/SEHC.2009.5069606.

he seldom directly collaborated with designers. Instead, he worked with a project manager, who shared his summaries and information with the design engineers for interpretation.

Ideas are left open to interpretation if designers don't work with the cross-functional team. Information also gets lost in translation from customer needs to design inputs. We want to remove the buffers and obstacles between design and our cross-functional team.

As much as the members of our cross-functional team are part of the project, they do have their particular goals, and they focus on their own part of the project or product. We must recognize these potential conflicts. In the end, what we aim for in concept development are design inputs that will fuel the creation of a great design. Our goal as product designers is to give them opportunities to talk together and with us so we can gather design inputs.

Work with the team using a facilitated meeting.

To co-work with a team means facilitating meetings. We need to work with our team on ideas, potential design inputs, and their priorities based on things that matter. We're talking about product concepts. Remember, at this point in our project, we've identified a need. There's a gap in the market, a potential new product, and we have nothing developed yet.

What we can do with our cross-functional team is facilitate meetings as directed co-work. We work toward a common goal: something not as big as the project goal, but something smaller and more focused that can be achievable within one working session.

During this co-work, we want to focus on the user. We're designing and developing something for potential users. That is a starting point.

We don't show up empty-handed at a cross-functional collaboration meeting. We want to do some homework since we need a roadmap to follow. It will provide a general direction for our conversations, prompts for group questions, and creative scenarios.

Directed co-work is not designing something together with a team. I am talking about exploring the use space and concept space with our team for design inputs. I'm not talking about design sprints, whether the Agile kind or product development sprints when you develop something in a week. We also want to avoid typical brainstorming sessions. In directed co-work, there are elements of creativity and some brainstorming techniques, but they reside within a structure.

Our structured meetings are going to give us an opportunity to talk with our teammates. They help us better understand their perspectives. Because we made them part of the design process, it will be easier to agree on design ideas we create later. We also avoid surprises because we've verified, checked, and left room to explore the use space. We've exposed different interpretations and hidden problems. These are things that, if done alone, we wouldn't have noticed. But together we figured it out.

The ultimate focus of this co-work is concept development. Although we want to direct that development toward design inputs, the rest of the team also gets inputs they need. Manufacturing gains more information about what they may need for their plan. Marketing gains more clarity about their plans. Reliability gets an improved understanding of the use space, environment, and its limitations. Quality gets an idea of what quality measures may be important, and project management sees where there could be project risks.

Everyone benefits, but as product designers, it's our responsibility to probe the cross-functional team for concept development information because they are our internal customers.

Collaboration is key.

Designers need others because they have internal customers and are at least one degree removed from the need. Effective collaboration and input from other individuals are crucial.

Early in my engineering career, my manager invited me to join an Operational Excellence initiative. The goal was to revamp the Quality

Management System for important parts of the business. Project leaders formed teams from different departments, supervised by a senior manager and assisted by a consultant.

The main idea was to comply with regulations, develop a compliant process, and exclude the current experts from the process and discussions about the new systems.

I'm not sure why the project leaders chose to exclude the current experts. Perhaps they thought the experts would carry over poor practices or challenge doing something a new way. It could also have been to jostle loose old ideas to bring in new ones. Maybe it was an attempt to overhaul a system that, over time, became disjointed and not streamlined.

All systems were affected, so it was not that they singled out any one group or system. Because it affected everyone, this project ended up as a significant source of angst for many in the organization.

"Imagine being a proud manager of a group, being told that other people, with little to no experience doing the thing you do, were going to create a brand-new system. And that you and your team had to follow and comply with this new system," the current manager of the auditing group confided in me. "And on top of that, you had absolutely no input."

My friend was nervous about what we were doing since nothing was communicated to him. He wasn't asked about their existing practices, which were based on relevant needs.

He felt he was at the whim of a stranger who decided what his group and the company needed. In our conversations over time, he went from feeling shocked to dismayed to angry and then defiant.

"I'll just change it back to the way it was after this project," he said.

In actuality, Operational Excellence created problems for his group. The auditing schedule was more randomized, but it also lost the measure of risk. As a result, we piloted the changes by going to a major pharmaceutical provider who was a well-established vendor with no prior issues.

"Why are we auditing this company?" the auditors asked. The risk was low.

Other aspects of Operational Excellence remained in place, such as continuing to integrate auditing more with other systems and formalizing other existing procedures. Managers and auditors recognized and appreciated the results of the Operational Excellence team, but the plan was not foolproof.

I'm sure they had reasons for structuring the project the way they did. But from my friend's point of view, it didn't seem right. He was really the internal customer, and we ignored our customer during development.

Usually, the person who identifies a design concept is someone familiar with the use space. They see a gap and an opportunity to fill that gap with something new. In the industry, these people are usually field technicians, field trainers, customer support roles, sales, and marketing. They are in touch with customers and find opportunities. New projects can also start with collaborations between companies or between a university and a manufacturer, for example.

The people who notice the problem or gap are usually not the designers. The designers are at least one degree removed from the problem, gap, and vision of opportunity. I'll call this "field vision." There can be more degrees of removal depending on how the team is structured.

A common example is when the project manager is an engineer and acts as the middleman between the field vision and the engineers. The greater the distance between those with field vision and the ones making daily decisions about a design, the more things fall into cracks or get lost in translation.

For those with field vision, handing off an opportunity to a team for development can turn into a situation like the one in which my friend found himself. He knows all about the problems, gaps, and sees opportunity. But it has been handed off to another group of people who do not know the field vision.

My friend isn't being asked for input or questioned about consequences. Whatever is developed, he's likely going to be stuck with trying to use it to fill the opportunity he saw, and it may not be a good fit. How many engineering concepts have you revealed only to face opposition or get a list of major changes?

Product development processes now include feedback loops for customer input. Teams approach customers and ask for input during development. This helps incorporate field vision into the process.

To further avoid this situation and improve the results of our efforts, the designers and the broader team members who have field vision need to have a shared understanding of the concept product.

Check your mindset to ensure success.

Now we've decided we need to get the cross-functional team or a team of people together to help us solve a problem, prioritize things, and gather information for design.

Just scheduling meetings can be difficult because everyone is already so busy, so we're careful about whom we invite and who has the information we need.

We sometimes run into a challenge where people say, "No, I can't make that meeting" or "This is not a priority for me." They've decided that whatever you need to talk about isn't worth the investment of their time.

Meanwhile, you feel like they need to contribute to whatever decision is being made or have important contributions to make. They may also need to attend to gain a better understanding of what's being discussed.

What actions can you take now to build goodwill for tomorrow?

People want to contribute their ideas, especially if it's going to help create new products that other people will love. But they also need to protect their time. We're inviting them to a meeting, and they'll assess whether they want to come based on that invitation. How should our team prioritize our request?

What has been our teammates' experiences in previous meetings? Some things they may consider are:

- Did it add value to me or the project?
- Did I walk away from our exercise with a better understanding of something?
- Was I respected and given a voice during the meeting?
- Was my opinion considered? Or was it just lip service?
- And what happened to the information? Did it affect the design, or did it just sort of fizzle out and disappear?

If their experience with us has missed the mark, they may now say, "No thanks," because now it's no longer a priority to take part in our co-working sessions. It's a shame for us because now we'll miss an important viewpoint and source of design inputs.

Be genuinely interested in what they offer.

The tools and steps I share in this book are to help ease some of the stress in working with others. It gives you ways to approach intangible concepts, collect data, and prioritize it to create a meaningful design.

The other part of successfully collaborating with others comes down to your mindset. You must genuinely be interested in other people during these collaborative meetings. Be curious about what they think, their point of view, and how their background leads them to believe the things they do. At the same time, try to stop yourself from designing solutions in the concept space.

In his bestselling book *How to Win Friends and Influence People,* Dale Carnegie tells a story of a successful magician and his way of thinking about his work. Imagine yourself as a magician, he says. You rehearse the show and prepare for it. You sell tickets, and your audience is waiting. What are you thinking to yourself just before you step onto the stage?

You might think: "I'm going to go out and trick and fool these silly people, and they'll laugh. Then they'll tell their friends about it, and I'll get more tickets for the next night's show!"

Or you think: "I love my audience."

That was the mantra of Howard Thurston, a famous magician, in the late 1800s to the early 1900s. He was an international star with more than 60 million people seeing his performances. "I love my audience," he chanted out loud before every performance. And they could probably tell. He attributed at least some of his success to this authentic mindset.[12]

How does this relate to collaborating with others? If you arrange and schedule these meetings with your cross-functional team, you must genuinely value their input and make space for their ideas. They will sense if you think their input is important and valued or not. If you just go through the motions of a meeting without really listening or staying curious, they are likely to perceive it as a waste of time.

Stop yourself from designing solutions too early.

Part of holding space for your teammates' ideas is stopping yourself from designing solutions during the meeting. Designing too early during these meetings is like trying to solve someone's problem when you're really just being asked to listen.

In his book *The Advice Trap: Be Humble, Stay Curious & Change the Way You Lead Forever*, Michael Bungay Stanier introduces the advice monster. It exists within all of us and desires to solve other people's problems. His book focuses on coaching habits, but it also applies to facilitating knowledge-sharing meetings with our teams.[13]

---

[12] Dale Carnegie, *How to Win Friends and Influence People: Updated for the Next Generation of Leaders* (Simon & Schuster, 2022), 57.

[13] Michael Bungay Stanier, *The Advice Trap: Be Humble, Stay Curious and Change the Way You Lead Forever* (Box of Crayons Press, 2020), 5–9.

When we jump into design mode during concept development, it's our advice monster rearing its head. It's an issue because we may solve the wrong problem or propose a mediocre solution because we don't fully understand the problem. We can't design great solutions until we fully grasp the problem. That's the core of concept development.

According to Stanier, jumping into solutions does several more damaging things. It demotivates the recipients of the advice, overwhelms the giver, compromises team effectiveness, and limits organizational change.

Here, we're asking the cross-functional team for input so we can solve a problem together. Open your mind to the ideas your team shares with you. If you get an idea, make a note on the side to come back to it later. You'll gather so much information from your team that whatever grand idea popped into your head will probably be obsolete after you take in all the inputs, anyway.

Good meetings can increase productivity.

Not starting the design process now takes patience. People are concerned about ruining their individual productivity, the team's productivity, and the project timeline. People are also tired of meetings because they interrupt their work. Meetings sometimes don't add value, and they can be tiring. It seems like simple math: Eliminate meetings and improve productivity. However, we need to take the time to actually talk with the cross-functional team. We need meetings.

A commonly cited figure on productivity is that a human being can expect to have three to five hours' worth of focused productivity in a day.[14]

Say we work in a nine-to-five job, which is eight hours a day. We have three hours of focused, productive time. That leaves us five more hours.

Let's give ourselves two hours for lunch and breaks. So now, we have three hours left. Let's say a good productive meeting with our

---

[14] Cal Newport, *Deep Work: Rules for Focused Success in a Distracted World* (Grand Central Publishing, 2016).

cross-functional team takes about one hour. Even after that, we still have two hours left in our day, which means that during a regular day, we can make time for a co-work session with our team.

8 hr. - 3 hr. productive work - 2 hr. lunch/break - 1 hr. co-work

= 2 hr. left in the day

Now, consider the productivity of individual work versus teamwork. Each of us has a time of day when we're most productive. Some productivity gurus ask us, "For you, what time of day are you most productive?" and then challenge us to sit down in a chair and get our work done during that time of the day.

When I'm focused on engineering or technical work, I source my productivity from a pool of my own internal energy. When I'm working with a team, my productivity energy source is different. I'm sort of getting charged by the energy being given off by others, even if it's on a phone call.

We can look at that productive cross-functional team meeting as extending our productivity time instead of limiting it by an hour. We also save some of our personal energy because our team shares the cognitive load.

3 hrs. productive work + 1 hr. co-work = 4 hrs. productive work

This assumes our meetings with our cross-functional team are actually productive. Our meetings need to have a purpose, an agenda, and a facilitator. If it's missing these elements, it's not a meeting but a party or a social gathering. If you follow the frameworks and process for team meetings that are in this book, then you're the facilitator who provides the tools toward meaningful teamwork.

Handle your teammates with the same care and respect you would show customers.

To reduce friction with our cross-functional team, it's important to treat our teammates as we would treat our customers. This applies to the co-working session, invitations, and follow-up. If you're leading the session and driving the analysis or information gathering, then pretend they are your customers for that session.

We have these kinds of meetings with our actual business customers. They're the ones who will potentially use our new product. We wouldn't approach a meeting with our customer with a casual attitude. We would want to put our best foot forward. We would plan and be respectful of their time.

It's that sort of care and attention our cross-functional team deserves. When we take that care and attention, they will want to take part more in co-working sessions with us, and the information flow is going to become easier and more abundant. In other words, our mindset should be:

"My cross-functional teammates are my customers. I'm asking them for information, and we're going to be making some decisions about it together."

What does this mean when it comes to planning a facilitated meeting? It's really about being organized and respectful. For one, we want to be prepared with the information that our cross-functional teammates may want to see as far as the scope and background of the meeting, or whatever topic we'll discuss. We want to have supplies and a plan for how we're going to co-work. By facilitating and guiding the team, we will own the meeting.

We want to make it easy for everyone to participate. That means the meeting should start on time and end on time or end early. This is part of respecting other people's schedules.

We want to allow time for proper co-work session closure. We don't want to use the whole meeting time for information gathering. We need to leave a little time at the end for the usual teamwork stuff. Closing the meeting includes agreeing on next steps and future actions, allowing time to collect notes before you have to give up the meeting space, and conducting any kind of meeting evaluations.

Follow up and follow through to show your team how their work developed into the solution.

If we're asking other people for ideas and information toward future decisions, then we're going to take ownership of the information they gave us. That doesn't mean we need to adopt every idea, but we need to let them know what happened next. If we do analysis and end up with recommended actions and next steps, we need to follow up and follow through.

As a quality practitioner, I facilitated and led a team through a failure mode and effects analysis (FMEA). The team created and recommended actions everybody bought into and wanted to implement. The participants worked together enthusiastically, excited about the changes for improvement. We assigned actions to a responsible teammate and set a due date.

Guess what happened next? Nothing. Things didn't get done.

I own that because I was the facilitator who led the group, and I should have followed up and followed through on those recommended actions with the right people.

I should have ensured that, after the meeting, they still agreed our recommended actions were important and do-able. If they faced resistance because of other job priorities, I could have followed up with their managers to ensure these recommendations were being prioritized. If it was a project we truly wanted to pursue, I could have worked on getting agreement from their manager.

Our team expects the co-work session leader to follow up and follow through on the ideas and information they shared. We may think that if it's their idea, or they're assigned to the action, they're solely responsible for getting it done.

But here's the thing: We can't control other people. We can only control what we do for ourselves as part of working with a group, so we need to own the follow up and follow through and maintain trust and good working relationships with our teammates.

There are many ways we do this while still holding our teammates accountable and not assuming their responsibilities. It could be a five-minute conversation to check in with the person assigned to learn about the status of an action.

It is acceptable to follow through by saying, "We prioritized this idea against the other needs of the project, and we will not implement this feature." Other forms of follow up they'll be happier about is, "Hey, this is how we're incorporating the idea," even if it's different from the initial concept your team devised. They all know design is iterative, so be sure to show them how their idea was developed and added to the final product design.

The actor Arnold Schwarzenegger provides a good example. In his self-help book *Be Useful: Seven Tools for Life*, he discusses the importance of follow up and follow through.

During his tenure as California governor, Schwarzenegger and his team developed emergency response processes and systems in the case of an environmental crisis. They practiced and felt they were prepared. Then the 2007 Southern California wildfires struck.

Here is where action matters. Some people assume everything's going to go as planned. But Schwarzenegger knew that's not necessarily the case.

Even though he kept in touch with people responding to the crisis, he decided he better show up in person with his team to make sure everything was on track. The team discovered problems, including signals getting crossed—and they stayed to coordinate the response on site so the things that needed to be done got done.[15]

We don't have to be state leaders to experience this kind of thing, especially in engineering and product design, where we're exposed to it all the time.

If it's a new product development project, we know development is not a linear process. There are going to be unexpected findings, failures, and

---

[15]   Arnold Schwarzenegger, *Be Useful: Seven Tools for Life* (Penguin Press, 2023).

lessons learned that we need to address. In early concept development, we also need to collect and prioritize many ideas.

If you're part of designing products and creating them, you have an important responsibility to many people. You have a responsibility to the customers to develop products that are safe and perform how we say they're going to perform. It's our responsibility to the business to create something the market's going to want to buy. We may also have responsibility for sustainability, both for the environment and for the people who work on the product, to help these products succeed and keep jobs intact.

We also need to answer to the rest of our team on the direction and decisions that affect the design itself. As Schwarzenegger says, "If you have a job to do or a goal you're trying to achieve, or you've made a commitment to protect something or someone and it's important to you that everything happens the way it's supposed to, it's up to you to follow through all the way."[16]

As a team, we may need to discuss this again and again to make sure we're all on the same page, but that is part of following up and following through.

During my career, I served as a quality engineer contracted to work on a new product development project. The primary quality engineer on that project had a whiteboard in his office. Anytime anything quality related to the project, whether it was the quality of the product itself or issues of regulatory compliance or the requirements of our design control procedures, it would get added to the whiteboard.

We continually addressed problems and repeatedly asked for updates. We didn't lose track of them, and we were able to see whether they were connected. It helped the team focus and follow-through on making the design better.

---

[16] Schwarzenegger, *Be Useful*, 93.

Mindset and consistency take us far.

With directed co-work, we handle ideas systematically with our team to get the maximum benefit from our creative phase. Here are some mindset guidelines for facilitating this type of teamwork:

- Recognize that it's difficult to evaluate ideas from group activities into actions for next steps. Be patient with yourself and others, and focus on the process, not just the result.

- It is necessary to control our itch for a quick decision on the best idea, which can undermine creativity and innovative ideas.

- We aren't looking to eliminate ideas. Instead, we're looking to develop them into the best solution possible.

- At the end of a session, check your progress and what you're learning and don't be afraid to pivot, if you want or need to. If you didn't reach the goal you wanted, then the team can choose whether to continue toward the goal for the next session or change it. Try not to pivot during the session itself, however. Let ideas play out first and see where they go.

Consistency will encourage greater participation. That means we need to be prepared, make it easy for our teammates during the meeting, show respect for their time, and follow up and follow through.

If we do those things consistently, then people are going to trust us with their time. We're more likely to get, "Heck, yes!" when you send out invitations for those co-working sessions.

The tools in this book help you be consistent. The ADEPT Team Framework is a checklist and a repeatable process. It helps you prepare for the right steps and keep a meeting on track and purposeful.

Key Takeaways:

1. Cross-functional teams bring diverse perspectives and improve design inputs.

2. Facilitated meetings are key to gathering design inputs.

3. Directed co-work during concept development involves exploring the concept space, not designing solutions.

4. Team members need clear communication and respect for their time.

Reflection Questions:

1. In what ways can I ensure team members understand their value and role in concept development, and how can I better facilitate their knowledge sharing?

   _____

   _____

2. How can I make meetings more productive and respectful of everyone's time?

   _____

   _____

3. How did I feel about the last meeting invitation I received? Was the purpose and value of the meeting clear to me? What changes would I make?

   _____

   _____

Chapter Three

A SYSTEMS APPROACH TO USER EXPERIENCE

Clients do not buy 'things.' They buy the experiences that those 'things' are able to deliver.

– Cindy Barnes, Helen Blake, David Pinder, experts in value proposition design and customer experience

———⟨⟨⟨⟩———

THIS CHAPTER introduces the ADEPT Team Framework and Concept Space Model as tools for team collaboration. We show that the framework is a process, and the model is a tool to help facilitate teams in developing concepts.

We've explored when to do concept development by making it part of the problem space, not the solution space. And we have showed why co-working with the cross-functional team is beneficial. However, we still have the problem of actually developing product concepts with a team.

There are no physical objects, or even drawings, to talk about because we haven't designed a solution yet. We don't want to jump into design mode until we've really explored the concept space. We want to stay in search mode, where we're finding problems and defining the use space.

To do this, we use a systems approach, zooming out to focus on customer experiences and working toward goals to fulfill those experiences.

Develop concept ideas by using a systems approach to focus on potential customer experiences.

User experiences can get lost during concept development because we are focused on developing the product itself instead of solving our customers' problems. We can lose sight of the customers because we are thinking about product function, performance, and features. The challenge, then, is to focus on the user's interactions with potential product offerings.

A systems approach helps us develop concepts before we design details. There are many instances of using a systems approach in engineering. A system diagram offers a placeholder for the product that we will develop but focuses on its context and interactions within the broader system. It also helps us organize information for design inputs. We can consider any concept development process from a systems perspective.

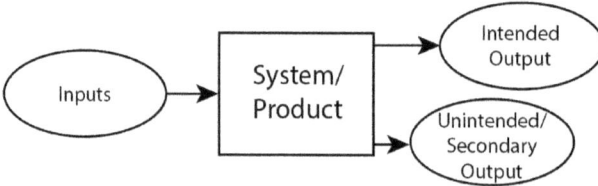

Figure 3.1. A Generic Systems Model

The center box within our systems diagram represents our product, which we haven't quite figured out. There is an input, and there are outputs. Inputs are activities or elements that are needed to make our product function or provide value. Outputs are what our product produces when it works. Outputs can be split into two types: intentional and unintentional.

Systems thinking works well for any product, whether it is static or dynamic. An example of a static product is a bike stand, which is unmoving. The input is a user loading their bicycle onto the stand. An intended output is a bike that is stored on the rack, secure and undamaged. An unintended output is an insecurely mounted bike that suffered damage when it fell from the rack.

A coffee machine with digital controls is an example of a dynamic, or functional, product. The input is the user adding filters, ground coffee, and water and choosing settings like brew strength and quantity. An intended output is a cup of coffee brewed the way the user likes it. An unintended output could be finding coffee grounds in the brewed coffee.

A systems approach for concept development helps us focus on the user experience, and we need to add the user experience to our systems diagram. At our inputs, we meet our customers where they are. They have certain expectations, understandings, and limitations.

Users are physically in a particular environment, choosing our product to help them do something, achieve a goal, or have a certain experience. They use our product. Then, in our outputs, our user experiences the benefits of our product when it works and the drawbacks when it doesn't. The **Concept Space Model** adds these user experiences to our development process.

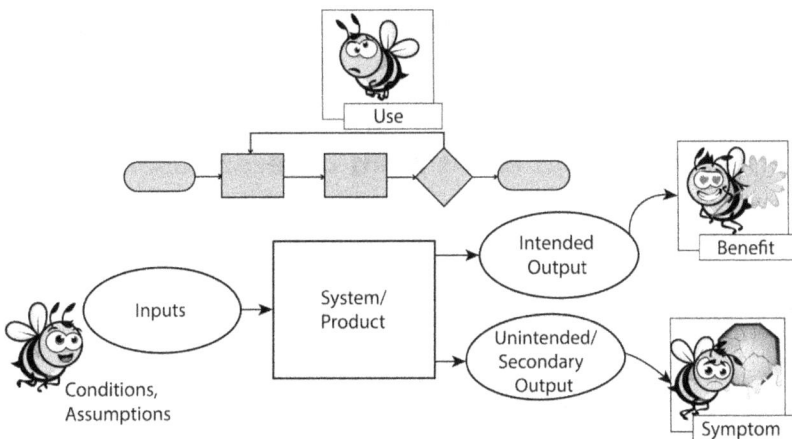

Figure 3.2. The Concept Space Model with Customer Avatars

When we zoom out to this higher level of product design, we're no longer focused on the design itself, but on the use space, made up of the customer experiences and their environment. The Concept Space Model lets us assess a product idea and its user impact.

Using a Concept Space Model with our team, we can achieve these goals:

- Improve our understanding of users and their needs by exploring benefits, symptoms (problems), and the use process, and better understand our customers at the input of our system model.

- Develop potential product features and offerings that will control outcomes and how customers use the product.

- Prioritize feature and offering ideas based on potential customer satisfaction, risks, and use steps that are critical to quality and add value.

When approaching any team activity, we need to balance having an end goal in mind and being flexible enough to allow for discussion and the growth of ideas. Before facilitating a meeting with your team, you need to be clear on your goals, and those goals need to be based on the type of information that is going to support your development of design inputs. You also want to ensure that, with the information you get, you can pull it through into the design so that you can show how your team's feedback is making it into the product design itself.

Generate concept ideas from different user experiences.

There are three areas of customer **Focus** within the Concept Space Model: Benefits, Symptoms, and the Use Process. Each of these relates to the Input. There are different concept design goals for each Focus. For each, we are improving our understanding, developing product ideas, and prioritizing those ideas.

Input Bee	Use Bee
I have a problem that your product will fix. You need to meet me where I am. You'll learn more about me through the other bees.	I am using the product. Make it easy and enjoyable for me. Reduce or eliminate mistakes I may make.

Benefit Bee	Symptom Bee
I experience the benefits from using the product. Some benefits I did not know about! I am loving what your product can do for me.	I experience symptoms of problems related to the product or its use. I may not be able to identify the problem. I am not happy!

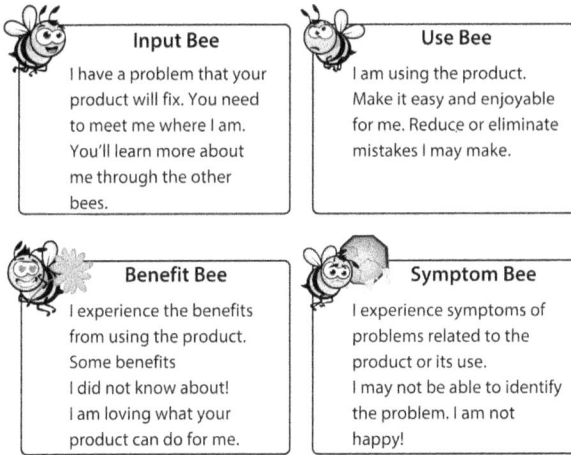

Figure 3.3 The Four Focuses with their avatars

With product design, we really want to meet our customers where they are in each of these areas. We can get important design inputs and prioritize them to make design decisions. From our systems model, we're working from the outside-in, starting with the user and then developing that into design inputs for design. There are many design inputs we can get from examining the concept space before we even start engineering solutions or features.

For each point of focus, we'll use bees as customer avatars. These bees represent and highlight that we're focused on our customers and their experiences, not on what our product does. Just as we use the Voice of the Customer for customer groups, we give our bees a voice. We have two customer avatars in our model output: the Benefit Bee and the Symptom Bee. We have another avatar during use of our product: the Use Bee. Finally, we are meeting our customers at our model input, represented by the Input Bee avatar.

The Benefit Bee is our customer avatar when things go right.

At our system output, when our product does the thing it's supposed to do, our customers experience a benefit of using our product. We want to target benefits and understand how much to implement them into the design. We want to maximize our design for benefits. When examining benefits, concept development goals are a list of potential benefits with ideas of features to create and options to increase customer satisfaction. The bee avatar would say things like, "I experience the benefits from using the product. Some benefits I did not know about! I am loving what your product can do for me."

The Symptom Bee is our customer avatar when things go wrong.

When our product doesn't work as we intended—or we had an unintended output—our users are experiencing the symptoms of a problem. Those are the events we want to minimize. With unintended outputs, we may design out or avoid them altogether. Concept development goals when evaluating symptoms are a list of potential symptoms with ideas of preventive features and risk controls. The Symptom Bee avatar would say, "I experience symptoms of problems related to the product or its use. I may not be able to identify the problem. I am not happy!"

The Use Bee is our customer avatar when they're using our product.

Our customers use our product to achieve a goal. They follow a procedure and take steps to get from where they start to where they finish, from input to output.

Each step can be an opportunity for our design choices. Effective design for use involves recognizing how use steps connect to the output and how inputs influence those steps. We evaluate what adds value and prioritize design decisions from that.

Concept development goals when evaluating the use process can be translated into a high-level use process flowchart, highlighted with important steps and use decisions. The Use Bee avatar would say, "I am using the product. Make it easy and enjoyable for me. Reduce or eliminate mistakes I may make."

The Input Bee is our customer avatar when they discover or choose our product.

Not least of all is the input to our concept space. The Input Bee customer avatar represents our customers before they use our product.

This is where we consider our assumptions and understanding of the use environment and our customers. Our customers also approach our product with their own assumptions and expectations. By examining the input from the systems viewpoint, we evaluate what we know about our customers and how they will use our product.

Through understanding the outputs and use areas of the concept space, we often need to change our inputs as we learn more about our customers. The Input Bee avatar would say things like, "I have a problem that your product will fix. You need to meet me where I am. You'll learn more about me through the other bees."

Generate ideas and prioritize each Focus differently.

By examining design concept ideas with a systems approach that is focused on user experiences, we're able to stay in the problem space and work with our cross-functional team to develop ideas.

Being able to prioritize the impact of our customers' experiences is a key feature of using the Concept Space Model. We're not just collecting ideas about the benefits, symptoms, and use steps. We are also discovering potential ways we can provide those customer experiences with our product. Plus, we can prioritize those experiences based on the best or worst case and their likelihood of happening.

Not all benefits are equal in the eyes of our customers.

With benefits, we can prioritize ideas based on a customer satisfaction rating. The higher the customer satisfaction rating, the more important it is for our concept design. We can extend prioritization to the impact it has on the customer and link that to how well we implement that into our design. Not all benefits are equal in the eyes of our customers.

Some typical questions we ask about the Benefits Focus:

- What are potential benefits when things go right?

- What features and offerings could provide these benefits?

- What options do we have to increase customer satisfaction?

- How well can we implement each idea into the design to achieve customer satisfaction?

Not all symptoms carry the same risk.

For symptoms that our customers experience, prioritization is related to the severity of the risk. The higher the risk, the more important it is for us to control or eliminate it, including through our concept design choices. We can refine our prioritization by considering not only risk severity, but also the probabilities of both the outcome and its impact. This helps us better understand the likelihood of the scenario.

At the Symptoms Focus stage, we should ask:

- What are potential symptoms when things go wrong?

- What unintended outputs do we need to prevent or control the most?

- How can we reduce, control, or prevent negative customer experiences?

- What are ways to reduce any negative impact on our customers?

Some use process steps are more important than others, depending on our criteria.

We prioritize use steps that are critical to quality or add value. By breaking out our customers' experiences and examining what is important to them and their successful use of our product, we can more easily prioritize design decisions.

Questions we can consider about the use process:

- What are the important functional steps and use decisions?

- Which steps are critical to quality and to achieve the desired output?

- Which steps do and do not add value?

- Do multiple people interact with the product to achieve the output, and if so, who does what?

For each customer focus, we have different questions to ask. For each of these customer experiences, we are improving our understanding, developing product ideas, and prioritizing those ideas.

Users and customers have individually targeted experiences.

So far, I have used the terms "user" and "customer" interchangeably. What if our user and customer are not the same? The customer is the one buying our product. The user is interacting with our product. These could be two different people. If so, then consider both of their experiences and needs in developing design inputs.

A Concept Space Model helps us develop ideas about ways to fulfill user needs and also focus on user experiences. Within this model, there is room enough to consider multiple user types.

Identify and track the Focuses that relate to a particular user or customer. Consider simplifying how you identify your customers to get the most out of developing ideas. At the very least, align the users and customers you evaluate with the needs of the project.

These are the goals you can expect when working with your cross-functional team within the Concept Space Model. For each of the Focuses (Benefits, Symptoms, and the Use Process), you can co-work to gain insight into better understanding, ideas for implementation, and prioritizing user experiences. Now that we know what type of information we're targeting with our co-work, we can choose some ways of working that help us and our team develop that information.

Convert wishes into actionable plans.

We've set some targeted goals for what we want our team to help us develop during concept development. However, a goal without a plan is just a wish.

GOAL - PLAN = WISH

We wish to promote meetings with our cross-functional team as long as they're meaningful. If we work with our cross-functional teams in a purposeful, meaningful way in a meeting, we can evaluate design inputs based on potential system failures or the things that could cause problems. To discuss the user and get more details about the use environment or better understand limitations in manufacturing, it's important to understand this is not information we email, slack, or text about. These things come up as we have discussions in meetings.

Right now, we're just wishing we could better work with our cross-functional team for design inputs. We're going to change these wishes by adding a plan to make it a goal. Our plan is going to include "how." We're going to get there by using models and templates as tools for cross-functional team meetings.

GOAL = WISH + PLAN

To explore different areas of the concept space, we will select a model and template for each. We will break it down into simpler processes to generate ideas. Additionally, we will facilitate communication and discussion of these ideas. In the end, we expect to gather valuable design inputs.

Key Takeaways:

1. Use a systems approach to think through different customer experiences.

2. Customer avatars help keep focus on the customers.

3. Targeting customer experiences helps prioritize work and design inputs.

Reflection Questions:

1. How often do I design from targeted customer experiences?

   _____

   _____

2. When considering the user's experience, do I give equal attention to benefits, potential symptoms to avoid, and the use process? Or do I tend to focus on one over the others, and why?

   _____

   _____

3. What other ways do I use a system model in my design? How is it different from the one in this chapter?

   _____

   _____

Chapter Four

THE ART OF USING MODELS TO IGNITE IDEAS

"It's really hard to design products by focus groups. A lot of times, people don't know what they want until you show it to them."

-Steve Jobs, the driving force behind Apple's era-defining innovations

———⊰※⊱———

THE FOCUS of THIS CHAPTER is how to use visual models and templates. We'll use them for structuring and scoping topics in early concept development. We also explain how to use them to facilitate idea collection and generation. We'll introduce you to the Benefit-Impact Model, Symptom-Impact Model, and Use Process Model. We use these models to identify features and inclusions for targeted customer experiences.

Concept development can be difficult partly because we don't have a baseline to consider for feedback. We don't yet have a product sketch in early concept development. It is difficult to explore and prioritize when we have nothing concrete to talk about.

There are risks in developing prototypes too early.

To get things started, many people create a prototype so they can get feedback and then iterate. Early prototyping has its place, and there are ways to do it well. For example, Design Sprints prototype the user interfaces of a design and then test with users. Reliability engineering HALT tests are used to test prototypes to failure to identify weaknesses. Bench top tests and simulations with prototypes help identify the feasibility of a design idea. In these examples, prototypes help answer specific questions.

Prototypes may not be suitable for early concept development when we don't yet have specific questions. They come at a cost that teams should account for when using them. Prototyping also takes resources to create, in both time and materials. Costs rise when we iterate on many prototypes.

There is another risk to early prototypes. Some teams, seeing the designers have already started, decide to forge ahead with the project. They are putting faith in the designers but robbing them of the opportunity for feedback. The team also loses the chance to have any real, significant input on the product design.

The team then develops a product built on many assumptions, whereas their original intent with prototypes was to explore and question those assumptions. They've lost an important step along the product development journey. We need models for a team to rally around that are faster and easier than prototyping and effective at developing design inputs.

By using visual models to represent ideas, you're giving yourself and your team power over creativity. We are more creative when we are limited with our options. Models can help by representing ideas and structuring conversations.

Typical brainstorming doesn't work.

Cross-functional teams struggle with effective collaboration. They often resort to standard brainstorming sessions to generate concept ideas. However, this approach does not yield desired outcomes.

Many people often see brainstorming as a chaotic process where all ideas are welcome and there are no restrictions. I have taken part in and facilitated my share of brainstorming sessions, and I was not thrilled with the results. I wondered if maybe I was doing it wrong.

Then I saw one of the chapter headings of the book *Inside the Box: A Proven System of Creativity and Breakthrough Results* by Drew Boyd and Jacob Goldenberg. It read, "How brainstorming produces fewer and lower quality ideas." This validated my brainstorming experiences.[17]

Alex Osborn, the founder of a U.S. advertising agency, created the creativity technique called "brainstorming" in 1953. The idea was to boost creativity by having teams of people work together with the premise that a group is more effective than an individual. Brainstorming became popular because it's a simple technique that can be used in many settings. Plus, the sessions can be enjoyable.[18]

During the 1980s and 1990s, scholars researched brainstorming to determine its effectiveness in problem-solving. They investigated factors such as the number of participants and the duration of the sessions. They also examined the true value of brainstorming and found that brainstorming is not advantageous for creative problem solving.

As the authors Boyd and Goldenberg concluded, "Brainstorming does not generate more creative ideas simply because people are in the same room."[19] Researchers found the brainstorming group possessed no

---

[17] Drew Boyd and Jacob Goldenberg, *Inside the Box: A Proven System of Creativity and Breakthrough Results* (Simon & Schuster, 2013), 30.
[18] Boyd and Goldenberg, *Inside the Box*.
[19] Boyd and Goldenberg, *Inside the Box*, 31.

advantages over the same number of individuals working alone. The group came up with fewer ideas that were lower quality and less creative.

Not having a way to work with a team is a significant challenge for concept development. We need to talk about intangible concept ideas. So, if not brainstorming, how can teams work together to generate ideas and creatively solve problems?

One study compared the performance of managerial groups who conducted a meeting using flip chart paper with groups of managers who used visual templates. The groups that used visual templates shared a significantly higher number of ideas, a greater range of ideas, and remembered more about the discussion, researchers found.[20]

Use visual models and templates for sharing knowledge with the cross-functional team.

A **visual model** represents a system, process, or concept. Models include diagrams and charts. They help people understand, analyze, and communicate complex information. Some models can be complicated, like three-dimensional models of an engineered product. For concept development, models represent ideas and will be simpler yet still meaningful.

A **visual template** is a structured framework designed to facilitate and guide the creative process within a team. It helps teams brainstorm, organize, and develop innovative ideas in a visually engaging and systematic way. A popular template is a Business Model Canvas. The Business Model Canvas is a tool used to plan and understand businesses. It's like a map that shows different parts of a business, such as what it offers, whom it sells to, and how it makes money. It shows how all these parts fit together so businesses can make decisions.

---

[20] Martin J. Eppler, Heidi Forbes Öste, and Sabrina Bresciani, ""An Experimental Evaluation on the Impact of Visual Facilitation Modes on Idea Generation in Teams,"" 17th International Conference on Information Visualisation, London, UK, 2013, 339-344, https://doi.org/10.1109/IV.2013.43.

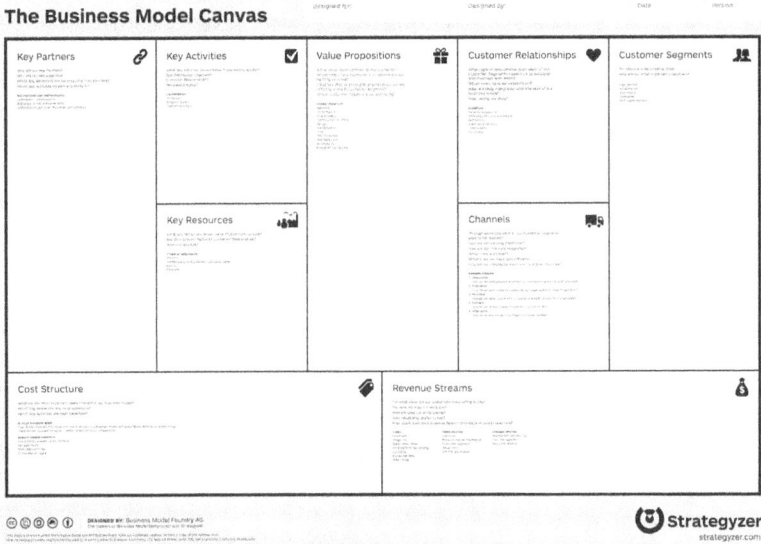

Figure 4.1. The Business Model Canvas: An Example Template
Image designed by Business Model Foundry AG (Wikimedia Commons, CC BY 3.0)

Visual models and templates are effective tools for representing and structuring ideas. They help narrow the scope of a topic. When used together during early concept development, they facilitate idea collection and generation within a team. This approach not only boosts the quantity of ideas generated but also enhances their diversity. Furthermore, using visual models and templates in discussions improves information retention.[21,22] Tracking ideas throughout the design process is crucial for incorporating them into the final product.

Visual models and templates are successful because of their association with Activity Theory. Activity Theory is a framework used to understand team activities in educational research, psychology, and organizational

---

[21] Marta Perez and Sabrina Brescaiani, "The Role of Visual Templates on Improving Teamwork Performance," *19th International Conference on Information Visualisation* (2015), http://dx.doi.org/10.1109/iV.2015.66; Martin J. Eppler, Heidi Forbes Öste, and Sabrina Bresciani, "An Experimental Evaluation on the Impact of Visual Facilitation Modes on Idea Generation in Teams," 17th International Conference on Information Visualisation, London, UK, 2013, 339-344, https://doi.org/10.1109/IV.2013.43.

studies. According to this theory, using shared and easy-to-understand tools improves teamwork, maintains focus, and prevents confusion. These tools or objects are clear, memorable, and ensure everyone focuses on the same things.[22]

Visual models and templates are helpful team tools to support our objectives in concept development. Their use can enhance idea generation and knowledge sharing in team settings. They provide a structured framework that helps teams focus their discussions and generate a wider range of ideas.

By using visualizations, we can overcome cognitive barriers during idea generation and knowledge sharing.

By externalizing thoughts and connections, visual templates make it easier for team members to build upon each other's ideas. This can be especially beneficial in early concept development when teams are exploring a wide range of possibilities.

As part of one of my jobs, I assessed the risks related to a medical device by talking to customers and field-vision experts. With medical devices, we're treating a sick patient. They are often intrusive and may be inserted into or connected to the body to provide a treatment. Sometimes, the medical device can cause harm.

In this case, we were evaluating a catheter insertion into a vein in the patient's forearm. One adverse event could lead to another, especially if the patient had certain health conditions.

A marketing representative for the device manufacturer helped set up meetings with a customer she knew so I could interview him. I met the

---

[22] Martin J. Eppler, Heidi Forbes Öste, and Sabrina Bresciani, "An Experimental Evaluation on the Impact of Visual Facilitation Modes on Idea Generation in Teams," 17th International Conference on Information Visualisation, London, UK, 2013, 339-344, https://doi.org/10.1109/IV.2013.43.

surgeon in his office and asked about product risks, rating them on a scale of 1 to 10. I also observed a procedure being performed.

The marketing representative provided feedback on product use, too, and we brought in a consultant who was an expert in the field. She had experience designing medical devices, was a licensed nurse, and understood standards. She was passionate about representing the patient's perspective.

"Hematoma is a risk. It's a severity of nine," she said during one of our meetings.

It became clear that her assessment of certain risks was much higher than that of the surgeon and marketing representative.

This was a challenging project for me since I was not well versed in clinical terms. Also, the interviews showed that our risk assessment process needed to be tweaked to improve it and introduce new ways of discussing complex situations. Once we'd done that, everyone clearly understood the risks and was confident not only in the assessment of risks but also in the methods to control them. We eventually achieved success and consensus.

What I introduced that helped us move forward was a common quality and reliability tool: a tree diagram. A tree diagram is a visual model of a chain of events, linked to show one or more chain of events.

By using this model, we could better come to a common understanding of the risk events involved in using our product. From our common understanding, we could better prioritize the risks.

Having a relaxed discussion with someone, like my interaction with the surgeon, is an effective method to explore a use environment or concept space. However, this sometimes gives you just ancillary or surface-level information.

Once you try to build out an idea with someone visually, like we did by using tree diagrams, a lot more information comes about. Such information can make a good design excellent. Or make a terrible idea into a fantastic one.

Integrating feedback loops during certain phases of product development helps throughout the development process. During concept development, we can use templates and models to discuss concepts.

The effectiveness of a visual template depends on its design and alignment with the task at hand.

Structured idea generation with visual models and templates improves idea quality. Researchers say visual templates help structure a process, improve focus, and lead to more and higher-quality ideas. When planning a meeting or workshop, it's important to choose or create visual tools that are right for the specific job. Generic tools may not work as well.

A multinational telecommunications company was seeking ideas for a five-year product pipeline. They asked employees to take part in unstructured brainstorming for six months using a generic visual template. The template prompted them to give their idea a headline, describe it "in a nutshell," and then sketch it. There was also space on the template for employees to list how their idea might relate to existing business platforms. Over those six months, people generated 93 ideas but only three were considered high-quality.

Then, company leaders changed their visual template to be more specific to innovation. The template was about the same size as the previous one but asked different questions with a single deadline. It still required an idea name and description, but it also asked employees to identify prospective customers, their motivations to use the product, and the benefits they would derive from it.

This time, the process resulted in 11 high quality ideas in a single session. The change from the generic template to the specific template represents a more than 300% improvement in the number of high-quality ideas. In addition, using the structured process with a visual template saved the company two months of work.[23]

---

[23] Perez and Brescaiani, "Role of Visual Templates."

Using a results table as a template doesn't work.

Sometimes, when we work towards a specific goal or endpoint, we may start from that endpoint and work backwards. For instance, if we need a comprehensive list of requirements for a project, we may involve our team in the process of populating a requirements table.

However, it is important to note that tools and reports like these are actually an analysis of information that was gathered during a separate process. This process involves actively working through ideas and gathering information with our team. Simply filling in the results to collect information isn't effective.

It's better to use visuals and templates that match the ideas we're working on. These models and templates can provide a structured framework for knowledge sharing and organizing information. This approach makes gathering ideas and information more efficient and productive.

Use the Concept Space Model to break down the whole.

We've already used a systems approach to break down the user's experiences, which is a start to defining a better scope for generating ideas. By doing this, we can focus on targeted benefits, symptoms to avoid, and our customer's use process.

Already, we have improved the type of design conversations we can have with our team. We've taken a customer need and broken it down into parts to better understand our customer and the problem they're having.

The Concept Space Model, however, is still too large for us to generate many ideas about a design concept. It serves as the foundation for generating ideas and defining the scope of a design concept. But it can still be overwhelming to extract specific design inputs that directly relate to customer experiences.

While the Concept Space Model offers an opportunity for alignment and a complete project understanding, we need even more focused ideas for a better customer journey. We want to delve deeper into design inputs that will enable us to create meaningful and impactful experiences for our customers.

So, we break down the concepts even more. We want to develop models and templates that target the benefits, symptoms, and the use process of our products, with the aim of creating design inputs that highlight our customer's experience. By creating these models and templates, we can more clearly understand the value our products bring to our customers.

This will enable us to effectively communicate the benefits, address any potential concerns or symptoms, and design for the use process. Prioritizing these design inputs helps align our concept with desired customer experiences, resulting in higher satisfaction and better outcomes.

Concept development models and templates help teams focus on users and design inputs.

"There is only one way to eat an elephant: one bite at a time."
—Aphorism

Imagine asking our teammates, "What is a list of potential benefits with feature ideas and options to increase customer satisfaction for our bike stand project?" They'd likely give us a blank stare, or they'd direct us to a report to find some answers. One issue is that the problem is too large to comprehend.

Another issue is that we're grouping too many thought processes together at once. To answer the potential benefits question, we need to use diverse skills and consider several different things:

- To list potential benefits, we are using our empathy muscle and putting ourselves into the position of our customers. We're also including thinking about their environment and competitor products.

- To provide some ideas of features involves using our creativity and thinking about how the customers use our product. Again, we're considering environment and competitor products. We also judge our ideas based on if they could cause our customers problems.

- For options to increase customer satisfaction, we're not just thinking about the product. Now we're considering the whole engine of customer-interfacing opportunities: customer service, packaging, color choices, availability, and more.

Asking our team to consider how to provide a benefit involves many thought processes and tasks. It is too difficult to try to develop all the design inputs to support this potential benefit in one discussion. It's asking our team to shift tasks too many times. This approach only considers benefits, ignoring potential symptoms and usage. Adding these other focuses compounds the difficulty.

This isn't a new concept. This is something you're probably doing daily or within your current projects to help you make progress. A way to help us with concept development is to break it down into smaller parts and smaller problems we think we can solve.

The Concept Space Model approach is the same approach applied to concept development, based on customer experiences.

Models organize ideas, create context, and help us understand the chain of events.

When we break things into their constituent parts, this helps us in several ways.

First, it helps us to dissect and organize ideas which can help us better understand what we're analyzing.

Second, it creates context. Context about a concept design helps us narrow our viewpoints to the things about it we can affect, which helps us toward meaningful actions.

Third, breaking apart events helps us explore the chain of events. Answering or at least considering why something happens helps us reflect on the factors that cause it. If we understand the things that lead to an event, then we better understand how to affect it. We also better understand the likelihood of the event itself.

Those are three reasons segmenting customer experiences is helpful: dissection and then organization of ideas, creating context, and understanding the chain of events.

Breakdown customer experiences even more for features and inclusions that lead to targeted customer experiences.

Figure 4.2. The Benefit-Impact Model

We break down each benefit into a potential available feature and the impact it has on the customer.

Our customers achieve benefits by using our product to complete a job, which then has a positive impact on them and their environment. By providing a product for our customers to use, we enable our customers to impact their environment.

The design inputs we desire from benefits are the features and inclusions we want our product to have and the ways we can provide targeted customer experiences. We prioritize design inputs that lead to these benefits based on customer satisfaction, balanced with how well we implement what leads to that satisfaction in our designs.

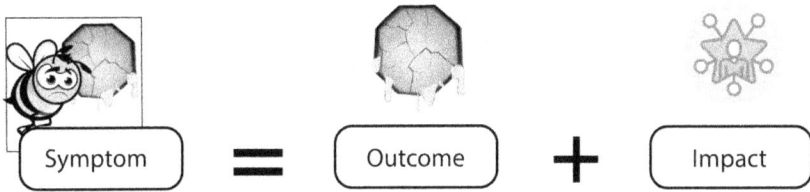

Figure 4.3. The Symptom-Impact Model

We also break down each symptom of a problem into the potential negative outcome when things go wrong, and its impact on the customer.

Our customers use our product to complete a job and achieve an outcome. Where benefits relate to intended outputs, symptoms relate to unintended outputs.

By using our product, our customers expect to make a certain impact, or at least not negatively affect their environment because they used our product. We examine symptoms to target customer experiences we want to avoid.

The design inputs we desire from symptoms are product inclusions, omissions, and other design inputs that will act to control, mitigate, or eliminate symptoms. We prioritize these design inputs based on the impact to the customer.

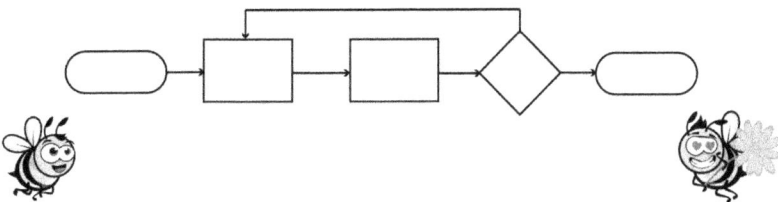

Figure 4.4. The Use Process Model

Our customers use our product by taking steps and making decisions. We can evaluate their use process with our product to get from A to B, from start to finish, mapping it out with a flowchart.

This helps align our team so we have the same understanding of the use process and work toward the same product expectations. We can perform other analyses to highlight things that are critical to quality or value added. By doing so, we highlight areas that are important for our design. We prioritize steps based on what affects benefits or symptoms, what is affected by inputs, and the value the use step adds.

We want to gather the important information and then prioritize it.

Part of knowledge sharing in concept development is identifying important pieces of information. Not all ideas are equal in the eyes of the customer. Because of that, not all ideas are equal in design.

If we are to design products focused on the user, we need to prioritize ideas based on users. We identify, to the best of our ability, what our customers will think about these benefits, symptoms, and use process steps. This not only helps us design products our users want, but it also helps us make decisions during the design process.

Prioritizing ideas based on users allows us to make informed trade-off decisions during the design process. We cannot include all features and functionalities because of constraints. By understanding what matters most to our users, we can make strategic decisions. This makes sure our product is great for customers and business.

The Concept Space Models break down the potential benefits, symptoms, and use process. Breaking down these concepts lets us further understand the impact on customers. Then, we link it to how users feel about their experience, based on our knowledge of them. Finally, we prioritize each of these factors based on how our users feel, using several methods.

We can create a rating scale that includes the voice of the customer data. Two-by-two charts are a way to map out the impact on customers. Our team can take part in voting techniques to choose between options.

We can compare use steps against quality measures and consider what is value-added to the user. We explore these in Part Two.

What we will get at the end of concept development using the Concept Space Models are:

- Better understanding of the use space
- Targeted benefits that are prioritized, with ideas of how to provide them and enhance their impact
- Targeted symptoms to avoid, prioritized, with ideas of how to avoid them and reduce their impact
- Use steps that are prioritized for quality and value

Knowledge sharing in concept development involves identifying the important pieces of information. By prioritizing ideas based on users, understanding their needs and preferences, and making informed trade-off decisions, we can design products that not only meet customer expectations but also drive business success.

Visual templates work, even though participants sometimes think they don't.

Part of trying new things with a team is sharing power over how their co-working sessions are run. When you use visual models and templates in concept development, you get a lot of information for design inputs. When you ask your team how they think it went, you may get feedback that is ambivalent, just warm to cool. They may have done the exercise but not think they got much out of it.

Although visual models and templates can improve results, participants may not recognize this. Their satisfaction levels could even be lower afterward.

Researchers thought this was curious, so they studied this aspect of visual templates, performing tests to show the improved outcomes while capturing participants' subjective opinions. What they found is that "subjects are systematically unable to correctly assess the benefits of visual templates they have been utilizing."[24]

To address this, we must emphasize the evidence-based benefits and highlight how visual approaches can improve concept development. Organizations should promote visual facilitation among team members, addressing any concerns about its value.

---

[24] Perez and Brescaiani, "Role of Visual Templates."

Key Takeaways:

1. Models and templates are productive tools for co-working with a team.

2. The Concept Space Model is a starting point to align on customers and their potential experiences.

3. Break down questions to get, and prioritize, design inputs.

Reflection Question:

1. Reflect on a time when you participated in a meeting that used a visual template. How did it help discussion with the group? Did it help the team reach their goal of the meeting? Why or why not?

_____

_____

Chapter Five

BEYOND BRAINSTORMING

"Before a decision-making meeting starts, be crystal clear about how the decision will be made."

-Bob Frisch and Cary Greene, Partners at the Strategic Offsites Group and Experts in Executive Team Strategy

———⟨※⟩———

THIS CHAPTER outlines the five key tips for successfully co-working with the cross-functional team. We also discuss stepping up to lead in concept development, which can take many forms.

Before we jump into using visual models and templates to coordinate idea generation among team members, we need a plan for how to facilitate co-work sessions. We want a consistent approach to achieve our concept development goals.

Define a scope with models and templates and provide a systematic approach.

Successful teams innovate by limiting the scope of work. They use models and templates to help them focus and ideate. And they use a systematic approach in the way they work.

One method to limit scope is the Closed World Principle, as outlined by Drew Boyd and Jacob Goldenberg in their book, *Inside the Box*. This principle limits us to things that are within our physical space and time. These are things already within our reach, sometimes right under our own noses and difficult to see. As the authors write, "the best and fastest way to innovate is to look at resources close at hand."[25]

They also use a Systematic Inventive Thinking process that uses templates, or patterns, to boost creative output. To use Systematic Inventive Thinking, teams create product ideas using what exists already, following specific patterns or techniques. Instead of thinking outside the box, they promote innovating inside the box. With the Closed World Principle and Systematic Inventive Thinking, they constrain the scope and define a systematic approach to get great results with team ideation.

In their book *Sprint: Solve Big Problems and Test New Ideas in Just Five Days,* Jake Knapp, John Zeratsky, and Braden Kowitz suggest a Sprint to test the user interfaces of a product for early input. They recommend limiting the scope of development to the user interfaces of the product.[26]

Their Sprint is a five-day process, starting with setting a goal and then ending on a user test with prototypes. A large part of the process involves a team iterating on hand-drawn sketches of ideas. As facilitators, they limit each iteration that team members create by number, space, and time. For Sprints, they limit teams to the user interface and follow a stepwise, structured exploration to develop ideas into prototypes to test.

Quality teams, like continuous improvement and Six Sigma teams, have also used models and templates for years to help them work together with success. Some of the models I've used include:

- Flowcharts

---

[25] Drew Boyd and Jacob Goldenberg, *Inside the Box: A Proven System of Creativity and Breakthrough Results* (Simon & Schuster, 2013), 9.

[26] Jake Knapp, John Zeratsky, and Braden Kowitz, *Sprint: Solve Big Problems and Test New Ideas in Just Five Days* (Simon & Schuster, 2016).

- Five-whys and cause-effect diagrams that help to get to the root of potential issues

- Fish bone diagrams to get everyone aligned in terms of potential causes of a problem, or solutions to a goal, and to help teams explore ideas and details

- Eight Disciplines methodology (8D), where teams use a template

These types of quality tools are visual models and templates that have countless stories of success for nearly 100 years across different disciplines.

In these scenarios, teams limit their scope of work to a target problem, working within a model or template. This allows them to focus and generate new ideas. How they ask the group to work is similar. They ask people to individually record their ideas, share and examine all ideas for common understanding, and then prioritize for decisions.

There are five key aspects to successful co-working with the cross-functional team.

What follows are some common ways people have been successful with team working meetings. We can adopt these techniques for ourselves and make things a little easier. We also can make it a more valuable meeting for our cross-functional team to share their ideas and for us to understand them. When we do it properly, we won't waste people's time while getting the information we need.

Add a boundary to the scope of work by focusing on one topic at a time.

Trying to take on all the details of a concept space in one session is too much to ask of a team and makes it difficult to share knowledge in a meaningful way. It's like we're asking our teammates to address several distinct areas of a product concept all at once.

Multitasking is the act of handling more than one task simultaneously. It is a common practice in modern life. But it decreases the effectiveness of idea generation because of cognitive bottlenecks, meaning there is only so much we can think about at one time. Multitasking is difficult because it can distract us with irrelevant information or stimuli that take us away from our goals.

People with high working memory capacity who focus on their work show more creativity and originality. This is important for coming up with creative ideas. Preventing distractions lets us systematically combine different aspects and options, resulting in more creative solutions. In other words, single-tasking with good focus beats multitasking for creativity.[27]

We can improve our working memory capacity through physical exercise, a healthy diet, and adequate sleep. We can also improve it by chunking information, which means breaking it down into smaller, manageable chunks. For concept development, we want this segmented approach for a deeper exploration of ideas. We need to allow people to sit with and think through a particular detail of the concept space. We ask one question at a time so they can focus and do a deep-dive into their knowledge base.

Limit the time teammates have to record their ideas.

The other activity that works is putting a little pressure on our teammates by limiting the time available to work. We need to give them some time to process what they know, but we don't need to give them *a lot* of time. Depending on the topic, we can ask them a targeted question and say they have up to 15 minutes to record their ideas.

We do this for several reasons. For one, we have other things to do in this knowledge-sharing exercise. Individuals sharing what they know is a

---

[27] Carsten K. W. De Dreu et al., "Working Memory Benefits Creative Insight, Musical Improvisation, and Original Ideation Through Maintained Task-Focused Attention." *Personality and Social Psychology Bulletin* 38, no. 5 (2012): 656–69, https://doi.org/10.1177/0146167211435795.

good thing. But we also want the group to understand the ideas and give feedback. While individual contributions are meaningful, it is the group's analysis of that information that should be a larger emphasis when it comes to both time and brainpower. Limiting the time to generate ideas keeps the co-work session on track.

Second, we focus better under pressure. Studies show that time limits can push people to come up with more ideas, faster during brain writing.[28] In brain writing, each person writes or draws their ideas on their own before sharing them with the group. When we need to decide something promptly, like choosing among ideas, we debate less with ourselves and just get it done.

For concept development, the focus is on sharing what it is we know, which requires less time than thinking about how many wild ideas we can come up with.

There will be a point where our team is furiously recording their ideas, then they'll reach a slowing point. It is then that they are trying to piece together an idea. Give them some time in that space but limit it.

Ask teammates to record their ideas concisely.

A common way to ask people to share their ideas is using Post-it® notes. We need to briefly describe our idea in large enough letters for our team to see on a little piece of paper. They're colorful, and we can stick them on a wall and move them around. They also limit the amount of space that we have to write.

Post-it notes are not the only way to limit idea descriptions. In Design Sprints, the team iterates hand-drawn sketches of ideas on a sheet of paper, using only one-sixth of the available space. Or, using online tools, people can contribute words to generate a word cloud.

---

[28] Lara Schmitt et al., "Dynamic Tabletop Interfaces for Increasing Creativity," *Computers in Human Behavior* 28, no. 5 (2012): 1892–1901, https://doi.org/10.1016/j.chb.2012.05.007.

Asking teammates to be concise when anonymously writing down their ideas helps with concept development in a few ways. First, it forces them to clarify and think through their idea, which encourages them to be succinct and specific. The more concise it is, the more real the idea becomes and the easier it is for the rest of the team to understand it.

Getting the team aligned with the same understanding of the shared ideas is an important activity. This not only improves knowledge sharing, but it also ensures everyone is prioritizing tasks based on the same understanding. It allows the ideas themselves to be the focus of attention.

We need ideas to stand on their own because we are choosing not to allow our teammates to have to defend ideas before the rest of the team.

Help teammates shift from "me" to "we": Record individually, share jointly.

Another important aspect of recording ideas is that our teammates do it individually before seeing what others have produced. It is their knowledge they're sharing. We want to give them some time and space to express their ideas.

At the same time, we don't want to give them an opportunity to compare their ideas against others. They may feel shy or conscious of office politics or other perceived limits. It also combats other common brainstorming issues, like managing forceful personalities.

Next, we need to ask our teammates to shift from "my" ideas to "our" ideas. The movement of physically sharing their idea on a common space, like a wall or a whiteboard, helps them see it in the public realm and shift their thinking from "me" to "we." Otherwise, people will protect their own ideas and fight for them, even if seeing their clever idea amongst other clever ideas is humbling.

Another way to eliminate competition or debate from knowledge-sharing sessions is to have everyone share their ideas in identical ways. Use Post-it notes and pens in the same color and make available a common wall or space (real or virtual) that everyone can access and read.

Since you are the facilitator, you are the "voice of the team" who summarizes and presents the ideas. If the team needs clarification, then ask a specific question about the idea for its author to answer.

Take care not to paraphrase too often, replacing your words with those of your teammate. Try to use their phrases and descriptions as they recorded them. The goal is for everyone to have the same understanding of the ideas, so that when the team decides what to do, everyone works from the same page.

Ask the team to prioritize what they've learned for design inputs and next steps.

Knowledge sharing in itself is adding value. Let's take it one step further and ask our team to prioritize what we've learned in this session and where the next focus should be. This is a way to focus on action and next steps, driving the knowledge deeper into the product development process.

Before we get to this point, we've already agreed as to how project decisions will be made. We can use priority scales as long as everyone on the team understands the levels in the same way. Multi-voting is a common way to decide the priority. For example, each teammate has three stickers to choose their three favorite ideas or portions of ideas. Or each teammate has one sticker to place on a ranked scale.

Reach a consensus on the decision, if possible. Consensus is that place where everyone supports the decision, even if it wasn't their first choice.

However, don't let anyone undermine the team's authority. An example of this is when a project manager tells a team they can decide, so the team chooses option 2. The project manager really thinks that option 3 is the way to go. So, the manager chooses option 3 for the project.

This kind of situation is demoralizing for teams, and they will be less likely to participate in the future. If the project manager wants to maintain respect and show respect for the team, then it is better that the team understands from the outset that they can provide options with their top pick, but the project manager will make the ultimate choice.

Stay on topic.

Sometimes the team will come up with ideas worth capturing but are outside the scope of the meeting. Don't dismiss these ideas. Just capture them in a "parking lot." A parking lot is a place to record ideas to come back to later or in a different meeting. The team should be able to see the parking lot during the meeting. At the end of the meeting, collaborate with the team on next steps for those ideas.

Being the leader takes many forms.

You can lead the team meetings and be the facilitator. As part of a cross-functional team involved in the design, this is the product you're designing, so you need to be confident in your capabilities to lead. That doesn't mean you can't ask for help. If you don't feel you're good at facilitating or want to share that responsibility with somebody, you can. Another thing some people find useful is meeting notes. If someone cannot fill that role, then I suggest you use AI to record and summarize the meeting for you.

While you're planning, keep in mind that you want to be hands-on to keep your team engaged in meeting the goal you have for that working session. You will need to set up a place for people to meet, including any kind of technology or other things you need, like a meeting space with a whiteboard and Post-it notes and markers. If you're not meeting in person, then you can use digital versions.

I've sat through meetings (and I'm embarrassed to say I've hosted meetings) where we're trying to fill in a spreadsheet during the meeting. Or we try to create a software flowchart as we go.

The problem is you're trying to make something pretty while you're working. This method just doesn't work for co-work in concept development. It slows down the momentum, which is the whole reason you're getting your team together.

Part of leading is also coordinating multiple schedules and inviting a diverse group of people. If you can, it's good to get a mix of experience levels from those familiar with past ways of doings things to those who are more likely to bring new ideas into the fold and challenge tradition.

Don't forget the decision maker. If you're creating a product where one or more individuals needs to make the final decision, make sure they are involved. They should either be part of the meeting or be able to review options you've narrowed down with your team. It is their responsibility to decide, so be sure they make it.

We all know if you get too many people in a room, or too few, it can be hard to get anything done. Some people say three to seven is the optimal number. I liked this point of reference my friend gave me: If you think about a meeting like a pizza party, you should only have to order one pizza. If you can't feed the people at your meeting with one pizza, then you have too many people.

Sometimes leading means rallying the team.

"Oh, well," lamented a coworker one late Friday afternoon. "We didn't make the shipment."

He reclined in his cubicle chair, hands locked behind his head, contemplating the view from his window.

"Is this the product the manufacturing team has been working hard on all week?" I asked. I was a process engineer and spent most of my time in the manufacturing area, so I was aware of what was happening.

"Yeah, but we'd have to get it all boxed up by 4:00pm. The rest of the team has already left."

I knew they had come in early to finish making product. It was 2 p.m.

"What's left to do to get it ready?" I asked.

"It needs to be packaged up."

"That's it?" It was an important step, but relatively insignificant compared to the work that went into making, testing, and inspecting the product.

I called over to my teammate. "Can you come to the basement with us?"

She stood up from her cubicle and looked over at me.

"We may need some help just packaging something," I say.

"Okay," she agreed. The other coworker raised his eyebrows and sat up.

I beckoned them to follow me. "C'mon, let's see what's going on."

In the basement manufacturing area, it was still brightly lit. A supervisor was still there, doing some paperwork.

"Hey," my coworker said, stepping closer to her workstation. "How much is left to do with the shipment?"

My other teammate and I ventured over to the packaging area. The manufacturing team had prepped product, lined up packaging materials, and made boxes.

The supervisor followed us. "These need to be boxed up, and I need to complete the packing lists and shipping labels," she said.

"Is this something you can do now? If we package these up, are you able to complete the shipping paperwork?" I asked.

"Sure!" she said. "You'll have to follow a certain procedure. The inspector is still here and can inspect it."

People had been working hard to make this shipment deadline, and it was an important one. To let it drop in the few remaining hours before the deadline seemed like a disservice to them. How would they have felt if they had done their part and the rest of the team didn't do theirs, especially when there was so little left to do?

"Let's do it!" I said.

The next Monday, we told the team the shipment went out on time, thanks to their contributions the week before. You can imagine their feeling of accomplishment. Leaders asked them to do their part to help meet a goal, and the rest of their team helped out, too.

It's not always best to swoop in at the last minute, but in this case, we weren't stepping on anyone's toes or doing someone else's work. It wasn't work that needed specialized training. And we weren't making it more difficult for anyone because we were there. Instead, we were pitching in as team members to help. I was told later it was amazing to see the "office people" in manufacturing, creating and filling boxes of product to meet a deadline.

When you see some members of your team struggling to get something done, help them by rallying the rest of the team to help complete the objective. And if someone offers their time to help you, even if it's collecting information, make it easy and productive.

Concept development often involves making sure the right people are in the room, talking together. It's about providing supplies or preparing models or templates to have a more meaningful discussion with a group. Starting and finishing on time also creates goodwill and shows respect for everyone's time, helping to foster good relationships at work and boost the team atmosphere.

Key Takeaways:

1. Brain writing manages typical challenges with groups and ensures anonymous idea generation. In brain writing, each person writes or draws their ideas individually. They aim to be concise and record as many ideas as possible in a short time before sharing with the group.

2. Consistent processes make meetings more productive.

Reflection Questions:

1. How can I better leverage the individual knowledge of team members by using brain writing?

   _____

   _____

2. How do I currently ensure team alignment on ideas? What new channels of communication would my team best respond to?

   _____

   _____

Chapter Six

BECOME ADEPT AT LEADING MEETINGS

"Individual commitment to a group effort – that is what makes a team work, a company work, a civilization work."

– Vince Lombardi, American National Football League coach

———

THIS CHAPTER is about how to lead meetings for concept development. We describe each step of the ADEPT Team Framework, from mindset to execution. Use this framework to help plan and lead successful co-work sessions in concept development.

Now we know what type of co-work meeting helps us be successful in concept development. If we think of each aspect on its own on the fly during a meeting, it could get messy. And we won't get the results we want. We won't achieve the knowledge sharing we need, and our team will be frustrated.

Pulling together these important aspects into a framework helps us avoid these problems. We want a repeatable process to help us be consistent yet also flexible to fit our needs. Think of this framework as a guidepost to planning and executing co-work sessions, whether you are the person planning the event or attending it.

Planners benefit from ADEPT by being able to think through and prepare for a co-work session. Attendees benefit because the facilitator is ready, and they know what to expect because it's consistent. When they see results, they are more likely to recognize the value of co-work sessions. In the future, they'll move their other meetings around to make time for co-work.

The ADEPT Team Framework includes steps to generate ideas and knowledge share with a team for concept development.

With any meeting or co-work session, we still want to put into place the normal practices for all meetings, like an agenda, next steps, and follow-up tasks.

There are three steps that differentiate a regular meeting from one focused on generating ideas and sharing knowledge with a team for concept development:

- Discover ideas: "What is it they know, and what are their ideas?"

- Examine those ideas: "Let's come to a common understanding."

- Prioritize those ideas to take action: "What decisions can we make to move forward?"

I arranged these ideas into an acronym that I use to help me plan useful meetings. I also use it during the meeting to keep me on track so I don't miss an important step and keep the goal of the future steps in mind. My system is ADEPT. When we get good at it, we'll be adept at facilitating co-work sessions.

Align
Discover
Examine
Prioritize
Teamwork

Figure 6.1. The ADEPT Team Framework
illustration by Freepik, adapted

If I'm planning a co-work meeting, I'm thinking about how to execute each step of the ADEPT process.

- What do I need to do to Align the team?

- How are we going to Discover ideas—with Post-it notes, virtual notes, or other list-building activities?

- How do I allow for everyone to Examine all the ideas created?

- What are our criteria to Prioritize our ideas against our goal? Do we have a rating scale or are we voting?

- Finally, what steps do I need to take to capture the results of our Teamwork?

During the team meeting, the success of ADEPT depends on us working toward the goals of each next step. To discover ideas, we need to align our team well. Once we've done that, we can examine the ideas to better understand and prioritize them. Then, to drive action through teamwork, we must discuss our priorities based on what we've learned.

Overall, we break down the scope and goal of our co-working sessions using the Concept Space Model. Each co-working session will cycle through an ADEPT process. We repeat ADEPT for each co-working session.

Typical Brainstorming ≠ ADEPT Methods

Typical Brainstorming	ADEPT Methods
• Free work	• Prework
• Expansion of ideas	• Prompted discovery
• Creative Prioritization	• Prioritization with criteria
• Unusual ideas	• Creativity with co-work
• Raises questions	• Drives Action

Figure 6.2. ADEPT Methods are not like brainstorming

A is for Align

With this first Align step, we want to bring everyone to a common starting place by explaining our scope and goals. We want to provide our team some perspective and review the problem we're trying to solve.

Meet your team where they are and then lead them to a common start.

During the meeting, consider the team's needs. They have not been preparing as much for this meeting as you have, so they do not know the details as well as you. They may also come from a different meeting or may have left a task unfinished at their desk to join you.

You need to help your team shift to co-working on this scope and goal. Consider yourself a tour guide: You need to gather and guide them to a point where everyone understands the task at hand and focuses on the same thing. The alignment step is getting the team ready to work toward a mutual goal together.

Figure 6.3. The first step is to align the team, like a guide
People illustration by storyset / Freepik.

Align your team by offering perspective, defining scope, and reviewing criteria.

Give your team some perspective on why they're getting together. Remind them the project they're working on and where you are in the development process.

Define the scope of what will be worked on during the meeting. Use the models to help you clearly show your scope and consider re-reviewing the models to get everybody warmed up.

Review the criteria you will use for prioritizing tasks or ideas. When planning for this meeting, you should have considered what prioritization is needed. If you are using a specific rating criteria or tables, review them with your team and ensure they have a common understanding.

Make goals visible to focus the team.

We always have a goal for each meeting that is clear and visible to the whole team. Goals for co-work sessions about the concept space could be: "A list of (number) potential benefits our customers experience when our product performs as we expect."

Or, in a later meeting, it could be: "A list of (number) potential benefits our customers experience when our product performs as we expect; described with potential features, services, and use environment facts that may drive them; and prioritized by customer satisfaction."

You do not need to make it as complex as a SMART goal because you are facilitating a meeting within a model. Consider the outputs of the model you're using to define a goal and keep it realistic in terms of what the team can achieve.

Communicate it clearly and make it visible. This goal should be on the agenda when you send it out. Then it should be at the top of the whiteboard in the meeting space or on whatever work surface you use. Try to keep it visible throughout the meeting. It will serve as a constant reminder of what you're trying to accomplish in this working session together.

D is for Discover

The Discover step of our team meeting is when we find out what our teammates know. We want to get their ideas out of their heads and in a place where the rest of the team can see them. We also want to group, define, or refine the ideas so we can Examine them in the next step.

When we get people together to share ideas, there can typically be some hang-ups. Some people are too shy or embarrassed to share in a group. Other people aren't shy or embarrassed at all and will be the only ones talking. Or you may have a team member who is overly pessimistic or optimistic or judgmental.

These are all things we can navigate so we can get everyone's input. Remember, diverse viewpoints make us most effective at generating design inputs.

Use brain writing to manage typical challenges with groups.

A method to help manage these challenges is to brain write. Brain writing means everyone writes, draws, or types their individual ideas in a shared space. It is a silent activity limited by scope and time, usually in short bursts of three, five, or 10 minutes at the discretion of the facilitator. Afterwards, the team shares their ideas anonymously.

Brain writing is an effective alternative to traditional brainstorming. It is widely used by Six Sigma and continuous improvement practitioners.[29] This method is also highlighted in the book *Sprint*, where team members draw out concept ideas for user interface designs. If this method is suitable for something as complex as user interface designs, then it can certainly be applied to listing design input ideas for benefits, symptoms, and the use process.

Check if the group has completed enough Discovery to Examine.

There is a key consideration you need to take as a facilitator, or the team needs to take together, between discovering ideas and examining ideas. We must ask ourselves, "With what we've discovered, can we discuss these ideas in order to clarify and better understand them?"

If the answer is yes, then let's move forward and examine them. If it's not—if it's just a bunch of disorganized ideas and we wouldn't even know where to start—then we're not done in the discovery phase yet. We need to stay in the discovery phase to further process the ideas until we're ready to examine them together and decide.

---

[29] "Brain Writing: Lean Six Sigma," Six Sigma Certification, accessed March 20, 2025, https://www.sixsigmacertificationcourse.com/glossary/brain-writing.

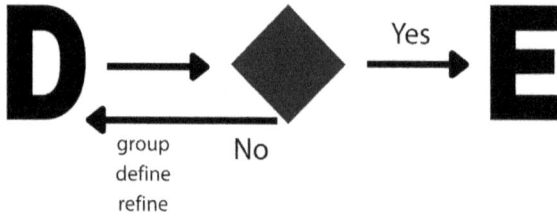

Figure 6.4. Decide if you can move on from the Discover phase.

Affinity Diagrams are a common method for grouping, defining, and refining information. People create these diagrams by moving pieces of information into groups. The groups can be predefined or determined as the team physically groups similar ideas together. There is no discussion—individuals simply move ideas around. If an idea fits into multiple groups, it can be copied onto a new Post-it note and added to another group. Eventually, all ideas are grouped.

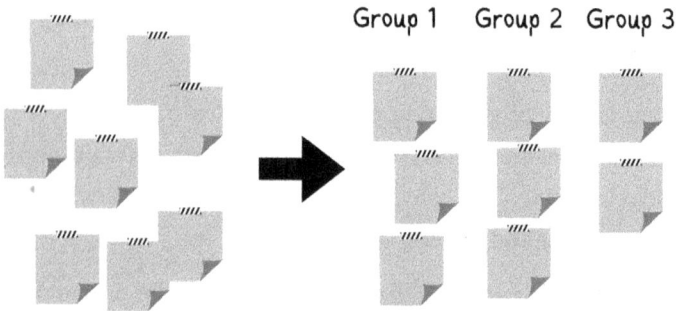

Figure 6.5. An Affinity Diagram

Here is an example of using an Affinity Diagram. You ask everybody to brain write, then they put their discoveries on the whiteboard or wall.

Now you have a mess of Post-it notes and lots of ideas. You're likely not going to meaningfully discuss or examine any of those things. You need to stay in the discovery phase longer and organize and group those ideas into subheadings that make sense. You're essentially creating an affinity diagram from your first brain writing step.

Once you've created your affinity diagram, the next step is for everybody to step back and view the whole. The facilitator then summarizes the ideas the team came up with. By having everything on display and summarizing the ideas, the team can then examine the ideas.

E is for Examine

After we've discovered ideas, we want to Examine them. During the Examine step of our team meetings, facilitators summarize the team's finding. The goal of this step is for team alignment on the ideas: Everyone is clear about each idea and has the same understanding. Our purpose in doing this is so everybody can assess the ideas and properly rank them in the Prioritize step.

The facilitator presents ideas so the team can perform the next step (Prioritize) based on the same understanding of an idea.

We don't want to have people presenting their ideas. If they do, that breaks down the methods we've built to combat those common brainstorming issues. Therefore, a facilitator presents the ideas the team has shared with each other. If the facilitator doesn't understand an idea, they can ask a specific question to the team to share an answer. Ask, get an answer, then move on. The facilitator should not cede their authority as the presenter of the ideas.

Take a "Yes, and..." approach.

We also don't want to have open debates or the too-rapid elimination of ideas. We avoid comments like, "That's a great idea, but here's why it won't work." We especially don't want, "Well, that's a dumb idea, we don't want it."

The mindset we want to encourage our team to have during this Examine step is toward understanding. We want them to think, "That's a great idea, what can we do to make it work?" or "What is it about this idea we can use?" We want to be forward-facing and considering possible actions through this process.

P is for Prioritize

After we've discovered what our team knows and all team members have examined the information, it's time to prioritize toward decision making. We need methods to collaboratively set priorities for early concept development.

In the Prioritize step, we make decisions based on our co-working goal and toward what is coming next. We look both backward and forward.

Figure 6.6. Prioritize against goals and what's next
vector designed by storyset / Freepik, edited

Prioritize options.

First, looking backward, we ask, "Considering where we started, did we do the work to reach our goal?" We assess if we've met those goals.

We think about the project overall and the goal we had for this meeting, a goal that is clearly visible for all to see. When we work with the Concept Space Model, we set ourselves goals like:

- List X number of benefit statements.

- List X number of symptom statements.

- Capture the high-level use process with start and end points.

Consider what the team has done in this co-work session. If they did not meet the goal, then we need to either choose more co-work to meet those goals or choose to adjust our goals.

Prioritize the next steps.

Second, we assess what we should prioritize for the next step of our work to move the project forward. We ask, "Considering where we are now, what do we need to do next?"

If we're further along in the concept space, we may have different goals:

- List drivers to benefits and assign a customer satisfaction rating to the impact.

- List drivers to symptoms and assign a risk rating to the impact.

- Identify use steps that are critical to quality.

For prioritizing at the Concept Space Model, our criteria could be to focus next on what's new, different, or unknown. Which benefits should we focus on in the next step, where we break them down for design inputs? Which symptoms should we focus on? What do we want to learn about our use process? Do we want to compare it, determine what is critical to quality, do a value-added analysis, or better understand multiple users?

Some analysis will have criteria built-in, especially later in concept development. For example, we assign a customer satisfaction rating to the benefit's impact, which is a priority. Because of that rating, certain potential features of our design will have a priority over others.

Get team consensus.

During our Examine phase, we discussed ideas to clarify them so everyone understands the same information. We need to ensure we all evaluate the same idea or the same understanding of an idea.

We'd also like consensus on an obvious option. Consensus is where everyone supports the decision, even if it wasn't their first choice. We don't need to pressure anyone to change their votes to force consensus. There are several ways to move toward consensus with your team.

Multi-Voting is a team method of choosing ideas. Each member silently identifies their top picks at the same time, using the same type of mark. For example, everyone gets five stickers to put on the ideas they like.

Decide the rules ahead of time:

- Put all marks on one idea.

- Place one mark each on an idea, choosing the top five.

- Stick one mark on one part of an idea.

A two-by-two chart is another way to prioritize ideas, like an urgent/important matrix. To make this method most helpful for your team, always keep "yes" as the most desirable priority, the thing people want the most.

At the end of the Prioritize step, we see which ideas have priority over the others. We'll see that an item has the most marks or a higher-level rating than the other items. Now we examine the results. Are the results split? If so, consider if the team needs more discussion to come to the same understanding of an idea.

T is for Teamwork

Teamwork is our last step in our ADEPT checklist for working meetings. We want our team to finish the meeting with results and action items. Ensure you have consensus on both. If your working meeting is to produce a recommendation, instead of collecting information for inputs, then get a consensus on one or two options with pros and cons. Remember that it's a decider's job to decide, so let them.

Next, ensure that you have captured the work so you can follow-up with the team afterwards. Take photographs or snapshots, whether in-person or virtual.

After you've closed the meeting, summarize the results of your team's work and share it with them and any other stakeholders. This may be a cleaned-up copy for a design file. Keep track of actions and ensure they are followed up on and completed. By sharing this with the team after each working session, they'll see how their work is aligning with the design inputs. They'll be better on board with the concept design and will likely continue to participate in these working meetings.

Structure each ADEPT cycle to complete the activities needed for that co-work session's goals and align goals to be consistent with the way we think.

Figure 6.7. For one goal with one prioritization criteria, complete one ADEPT cycle.

Usually, one ADEPT cycle is reserved for one meeting. There are times when we may want to go through a few ADEPT cycles or steps in one meeting. This depends on the scope of our work and how our team is doing with the activities.

It may also depend on people's schedules. If everyone is off site to focus on this one project, then you'll likely be doing multiple ADEPT cycles for concept development in one day.

For extended co-work sessions about various goals using different prioritization criteria, complete each goal with an ADEPT cycle.

I talk about using multiple ADEPT cycles more in chapter 7, where we use the Concept Space Model. With that model, we are focused on different goals about the Concept Space: list and prioritize benefits, do the same for symptoms, and create a high-level use process.

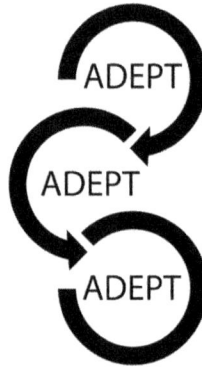

Figure 6.8. Multiple ADEPT cycles for goals with different prioritization criteria

All of these are distinct goals and ideas around the Concept Space. To complete each goal requires us to shift our thinking about the concept. In the end, we'll have three distinct sets of ideas and activities for follow-up. Because of this, I recommend using separate ADEPT cycles, even within the same co-work session or meeting.

To approach different goals, use one ADEPT cycle for each goal. Keep all steps of the ADEPT Team Framework in each cycle. Make sure to include the Align and Teamwork steps, as they can provide specific goals and actions.

For different items aligned to the same goal and using the same prioritize criteria, consider multiple Discover, Examine, and Prioritize loops.

Other times, we are in deep-thinking work and we'll use one ADEPT cycle with multiple Discover, Examine, and Prioritize loops. I recommend this approach if we're co-working with our team toward one goal but exploring multiple items toward that goal.

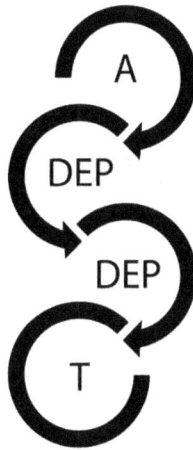

Figure 6.9. One ADEPT cycle with multiple D-E-P loops for one goal and prioritization criteria, but multiple items

I talk about this more in chapters 8 and 9, when we get into detailed work for benefits and symptoms. For example, we have a list of benefits we're going to explore further with a Benefit-Impact Model. We have one goal, which is to explore benefits. We have one set of prioritization criteria. We apply the same line of thinking to each, just to multiple items. In the end, we'll have a prioritized list of a distinct idea: benefits. We can also do this when exploring symptoms.

When approaching a co-work session with multiple items but one goal, I recommend using one ADEPT cycle. However, modify it to have different Discover-Example-Prioritize steps for each item. Ensure the multiple D-E-P cycles can be done with a consistent line of thinking by using the same criteria to assign priority.

For multiple prompts needed to prioritize, use multiple prompts in the Discover phase.

In some cases—for instance, process flowchart analysis—we are looking at multiple use steps. We're still focused on one goal and one prioritization process. We just work through several prompts in the Discover phase to compare for the rest of the ADEPT process. Just use one Discovery step for each prompt.

Figure 6.10. One ADEPT cycle with multiple Discovery prompts
to reach a point where the team can prioritize

These are all variations of the ADEPT model, only modified slightly to accommodate different types of information gathering and prioritization. In part 2, I include the appropriate symbol next to the Team Activities where the ADEPT Team Framework is used.

Key Takeaways:

1. Three steps—discover, examine, and prioritize—distinguish idea generation and team knowledge sharing in concept development.

2. The ADEPT Team Framework is a system for planning and executing co-work sessions.

3. Choose an ADEPT cycle for each goal and modify it for multiple items, if needed.

Reflection Questions:

1. How does the ADEPT Team Framework differ from my typical meeting process?

   _____

   _____

2. Am I consistently following up to ensure that team contributions lead to action and are incorporated into the design process?

   _____

   _____

Part One Review

By working through this review and activities, you are on your way to improving your concept development process.

Key Concepts

Listed below are some key concepts covered in part 1:

1. It's crucial to thoroughly define the problem before jumping to solutions. Projects are more successful when teams do solid "homework." This involves understanding the gap between the customer's current state and their desired state.

2. Designers should collaborate with a cross-functional team before creating solutions. This involves sharing knowledge and moving ideas toward prioritized design inputs.

3. We use a systems approach at concept development. This enhances our understanding of users and their needs, helps us develop potential product features and offerings, and prioritizes these ideas based on customer satisfaction, risks, and critical use steps.

4. Visual models and templates improve idea quality by structuring the process, improving focus, and leading to more high-quality ideas.

5. Consistent processes and a team approach are important to achieve success in concept development.

6. The ADEPT Team Framework is a process and the Concept Space Model is a tool. They are used to facilitate teams in developing concepts rather than designing solutions. They also help teams work together so they can identify and prioritize design inputs.

Practice Activities

1. Consider a project where the team jumped straight to a solution. List the challenges they experienced as a result. Overall, what did it cost? Consider the bigger picture like time, expenses, failed launches, unhappy customers, and reputations.

2. Identify specific strategies of when to incorporate a questioning or investigating phase for concept development into your next project. Does it correspond with a standard project phase or approval gate? What step should be paused before the team does this exploratory work?

3. Evaluate a recent project team for its cross-functional representation. Identify any gaps and plan to remedy them. List what you'll do.

4. Assess how you can show genuine interest in other people's ideas and encourage knowledge sharing rather than offering solutions. List three common "advice monster" responses you tend to use. Come up with three questions to use instead to elicit knowledge.

5. Categorize your customer information for a project to each Focus. Use the bee avatars to help you decide where it best fits. Are you applying most of your effort to one Focus? Why? What can you do to give each Focus attention in design?

6. Plan a team meeting using the ADEPT Team Framework, focusing on a specific design challenge. Include materials, location, and methods. Imagine facilitating the meeting and staying on track with the ADEPT Team Framework.

7. After your next team meeting, create a summary of the results of the work. Start connecting the results to next steps for design.

Part Two

PRACTICAL STRATEGIES FOR CO-CREATING POWERFUL DESIGN INPUTS

IN PART 2, we'll execute activities with our team. We'll co-work with them during concept development to share knowledge and help us define design inputs. In this part, we will:

- Put the ADEPT Team Framework and Concept Space Model to work, together, to get alignment on scope and targets.

- Turn our Benefit-Impact Model and Symptom-Impact Model into templates to collect information for design inputs.

- Use the ADEPT Team Framework together with our targeted templates.

- Review ways to prioritize information based on customer impact.

- Explain how process flowcharts and their common analyses can be used for prioritization in the use process.

Now that we have models and a method to co-work with our team, we're ready to execute. We want to use these methods to fully explore concepts between our problem statement and design inputs, or the dark areas of our double-diamond development process (see illustration). To achieve this, we use the Concept Space Model to fill in the first diamond, and we break down customer experiences in the second diamond.

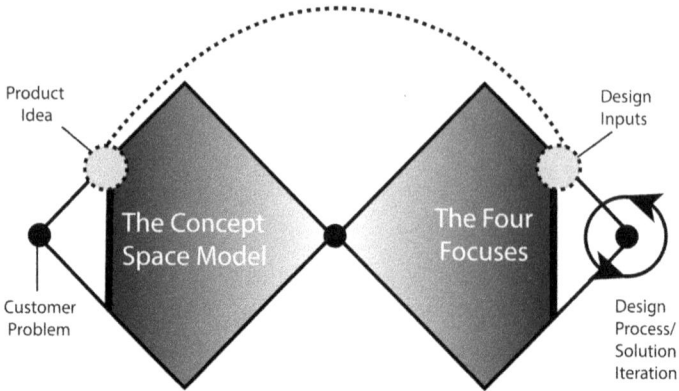

Figure P2.1. Where The Concept Space Model and The Four Focuses Fit within the Double Diamond Design Process

Host co-work sessions by goal.

We've discussed why breaking down ideas into chunks is a strategic move in concept development. We need to segment our meetings to match those chunks. In addition to design inputs, there may be additional editing to the concept space itself, such as changing the users or getting more detailed in the information. Plan ahead for short co-work sessions with a specific goal and be willing to cycle through them if your team feels the need.

The first goal is to explore the concept space at its edges.

The concept space is how the team gets alignment on users and the use environment, understanding where this new product will be used and what it will achieve. It's identifying customers and performing a gap analysis to understand what our product does for them. We identify a list of benefits, symptoms, and start and end points of a use process.

The second goal focuses on benefits and then explores them.

The team creates ideas for how a benefit may be achieved and elevated. They link the benefit's impact to customer satisfaction. Using customer satisfaction ratings, they prioritize benefits and their associated ideas. This joint information gives direction to the design team about potential design features that can be most important for customer satisfaction and how those features link to satisfaction.

The third goal focuses on symptoms of bad events and then explores them.

Symptoms are explored like benefits, except they are viewed from a risk perspective. The team creates ideas for how a symptom occurs and can worsen. They then link the symptom's impact with a severity rating. Teams may also assign a likelihood of the bad event, given that certain scenarios are more likely to occur than others.

They can rank the symptoms and related ideas by how serious and likely they are. This information gives direction to the design team about potential design features that can be most important to reduce risks and how those features link to negative effects.

The fourth goal focuses on the use process.

Teams map the steps customers take to use the product. The product is still a black box, but we can analyze the high-level process.

The team maps out the process steps to gain alignment and understanding. They then choose to analyze the process to prioritize the use steps. They may choose to identify the process steps that are critical to quality or adding value. Or they may break apart the process for multiple users.

These process analysis methods help the design team figure out inputs of the use process, which can drive decisions about the usability of the design.

We discussed some best practices in the first part of this book. In this part, we execute those strategies for concept development, leading to design inputs. Here are some reminders, which I've made statements of what *not to do* to provide a different perspective.

Things not to do when working with a team in concept development:

- Failing to explore the concept space early enough, missing information that is crucial for decision-making when it can have the most impact.

- Not choosing the team wisely, lacking essential experience and familiarity with the problem space. This results in a failure to gather the diverse perspectives needed for a complete vision of the product's performance and potential risks.

- Neglecting to focus on user experiences during concept development means the product design can become detached from what the customer truly wants and needs.

- Not preparing the team for co-working sessions results in unproductive and frustrating meetings.

- Failing to use targeted models and templates leads teams to struggle with generating and organizing ideas.

- Not using a consistent approach for team collaboration means team members cannot rely on a predictable process, potentially viewing meetings as a waste of time.

- Working in isolation or failing to ask for help when needed prevents designers from accessing the crucial knowledge held by cross-functional teammates.

- Not sharing what the team learned after each session breaks the feedback loop and prevents the team from seeing how their valuable input is used.

Chapter Seven
EXPLORE THE
CONCEPT SPACE

"The purpose of a team is not goal attainment but goal alignment."

—Tom DeMarco and Timothy Lister, authors of "Peopleware: Productive Projects and Teams" and leading voices in software engineering management

———◁◆▷———

IN THIS CHAPTER, we use the Concept Space Model and the ADEPT Team Framework to identify and choose targeted customer experiences. At the end of this exercise, we will have agreed on a description of our customers, and we have a list of benefits and symptoms to explore. We'll also have a high-level use process which will further define the scope of our work. Our first step in co-working with our team is to get alignment about why we are meeting and our goals. Our overall goal is to develop design inputs.

If leaders already vetted and approved the project, we should have a customer defined and at least some preliminary needs outlined, even if they are not formally stated in a needs document. Look to the project charter and market analysis for descriptions of the customers. It should be relatively easy to define the customer at the input of the Concept Space Model, at least at a preliminary level.

To get started on alignment, we introduce the Concept Space Model to our team.

Don't spend too much time at first trying to understand customers.

If you focus on trying to thoroughly understand the customer before you do anything else, you'll get stuck talking about them and not get to the scope, benefits, or symptoms.

By discussing the scope, benefits, and symptoms, you and your team actually gain a better understanding of your customers. This understanding enables you to clarify the specific needs and desires of your target customers. It helps you determine the scope of your project, ensuring you address the right problem and provide the most relevant solution.

Additionally, understanding benefits allow you to tailor your offerings to meet customer expectations. By examining the symptoms or challenges your customers face, you can identify pain points and develop strategies to alleviate them.

Expect to revisit your customers' descriptions, conditions, and assumptions regularly throughout the work on the Concept Space. As you progress in your project, new insights may emerge. You can further clarify and refine what you know about your customers as you explore the Concept Space. This iterative process of refining knowledge about your customers during concept development helps you deliver a more effective and customer-centric product or service.

In short, don't worry about nailing your customer descriptions before you start concept development. You'll learn more about them with your team through co-working with the Concept Space Model.

Check that your customers are also your users.

Is the person buying the product the same as the people using the product? They may be different people with distinct roles and interests. Dig into the market analysis a little deeper to analyze the target audiences and their specific needs. Understand the relationship between the buyers and users.

If it's not clear what needs belong to which group, then the time to work it out is now, during concept development. Use the Concept Space Model to help clarify those uncertainties. By identifying the distinct needs of both parties, we can develop a product that satisfies both.

The team may want to consider if it's worthwhile to have separate co-work sessions—one for the user and another for the buyer. Each session would have their own version of the Concept Space Model. If you do this, I recommend you first focus on the user. You may want to add a goal to that session: "Decide if we need a Concept Space Model for the buyer."

What develops from the user probably also applies to the buyer. After all, the buyer likely chooses a product the user wants. Afterwards, the team can decide if they want to explore the buyer with the Concept Space Model. You're doing this to share knowledge for design inputs. If a greater understanding of the buyer is going to lead to unique design inputs, then it's likely worthwhile.

Needs, benefits, and features are different things.

When we first start talking about customer experiences, our team is likely to confuse needs, benefits, and features. They are all related, but there is a way we can think about them that helps us develop design inputs.

A need is a gap in the problem space.

Our team has already identified needs, which prompted the whole project. Needs are gaps between what the user can do now, or has now, compared to what they want to do or have. They also identify and characterize the problem space.

One way of looking at needs is to consider the job to be done, which focuses on the outcome. It challenges us to question if the current way is really the best way of doing something. Another way of looking at needs is through user personas, which we employ to consider the reasons or motivations behind our user's behaviors.

Needs may be something our customers have identified. It also might be something they really haven't even considered.

Our project goal is all about developing a product our customers will use to fill a need. We're not only providing a device, but we're also enabling our customers to use it.

A needs statement can be worded like this:

[Our customer] needs [to do this] so they can [achieve this].

An example of a needs statement is this:

"[Rick, a busy biking enthusiast,] needs [to safely store his bike in his apartment] so he can [secure his bike from damage, prevent damage to walls, free up floor space in his apartment, and perform bike maintenance.]

You can see how both the task and the customer persona have input into the needs statement and how it describes the problem space.

A benefit describes the user's experience.

Benefits are positive outcomes of using a product that fill the needs gap and the impact it may have on the user. Experiences can be real or perceived. A benefit statement links products and users together. Whereas the needs statement describes the problem space, the benefits statement gets more specific about a targeted user experience.

A team could word a benefit statement like this:

> **[Our customers] can [use product capabilities or have characteristics] so they can [experience this value].**

An example of a benefits statement is this:

> "Users can [assemble the bike stand within minutes of receiving the box] so they can [use the product quickly, gain confidence in their decision, and have a sense of accomplishment.]"

Must users assemble the bike stand within minutes to fill the needs gap? No, assembling the bike within minutes is not a need. But it is an experience that will give our customer satisfaction. What if you have some needs that sound like benefits? Then, examine them as part of the benefits statement and include them in these exercises! The point of the exercises is to gain clarity on the design inputs.

Features are fuzzy parts of our concept product.

When our design project is complete, specific features will be tangible and measurable. This is not the case at the concept phase. Our product features are the product capabilities or characteristics we design for our customers to use.

Our team may discuss generic features to develop design inputs. However, we do not detail features or engineer them on the spot during a benefits statement exercise. We simply gather potential design inputs for concept development.

Facts	**B**eliefs or reasons
Essential function or component	**E**ffects of features
Attribute	**N**arrative
Test data describes it	**E**motional connection
Unique	**F**ocus is the user
Reason it works	**I**mproves a situation
Explain them with instructions	**T**ells how a feature adds value

Figure 7.1. Feature versus Benefit

Collect ideas and move on.

When we work with our team to develop ideas, we can get stuck in trying to categorize which ideas are really a need, a benefit, or a feature. I encourage you to collect the ideas and not worry too much about categorization. Because we're breaking down our benefits in the next step, it will become more apparent what is what.

Team Activity: Gather a list of benefits statements.

You may already have a list of benefits you want to explore. Analysis performed to approve the project may provide you with valuable benefit ideas to start. Examples of these are marketing analysis and initial customer surveys. Some user needs may really be a benefit, so you may start there, too.

If you don't have a list of benefits, consider developing them now according to the following steps. Your goal at the end of the working session is to have several benefits statements.

The steps to gather a list of benefits statements follows the ADEPT Team Framework and uses the Concept Space Model together. These steps take one ADEPT cycle:

Do pre-work. Gather customer descriptions and preliminary needs. Review the ADEPT Team Framework to get ready to lead the meeting.

1. **Align** on a scope and goal of the working meeting. Make the goal visible on agendas and during teamwork. An example: "A list of (number) potential benefits our customers experience when our product performs as we expect."

2. **Discover** potential benefits statements, using this template:

> [Our customers] can [use product capabilities or have characteristics] so they can [experience this value].

Brain write ideas and share them. Use affinity diagrams to group and refine ideas, so you can examine them in the next step.

3. **Examine** the list of benefits statements. A facilitator summarizes the group's findings for understanding and clarity.

4. **Prioritize** the benefits for further exploration for design inputs.

   4a. Multi-voting: Display the list of benefits. Each team member gets five dots (stickers, or they can draw them) and places one dot each on the benefits they choose to continue exploring for design inputs. They should not put more than one dot on any one benefit.

   4b. Review and discuss the results. Aim for consensus, which is that place where everyone supports the decision even if it wasn't their first choice. If you do not reach consensus, consider going back to the Examine step to ensure all team members are voting on the same idea or the same understanding of an idea.

5. Perform the **Teamwork** activities to close the meeting. List actions for follow-up, including items noted on the side in the "parking lot." Take pictures or save whiteboards.

6. Summarize this team design activity in writing and share what you've learned.

   6a. Summarize benefit statements and their priorities and share with the team.

   6b. Update user descriptions and their needs.

Once you've completed this ADEPT cycle, you'll have a list of benefits statements. You'll also have team agreement on which benefits to continue exploring for next steps.

Symptoms, failures, and hazards are different things.

Our team is also likely to confuse symptoms, failures, and hazards when we work within the concept space.

Symptoms are what customers experience when their product use results in an unintended output or event.

Our customers use our product to achieve a goal. There are the intended outputs, and there are also unintended outputs. In the Concept Space Model, a failure of some kind may cause this unintended output, such as a mistake during its use or something in the environment. Whatever the reason, it is a negative experience for our customers. We want to examine potential negative experiences so our design can eliminate, reduce, or otherwise keep them from happening, if possible.

A team could word a symptom statement like this:

> [Our customers] may [lose product features, be challenged] which leads to them [experiencing this negative impact].

We can translate some needs into symptoms by evaluating the opposite of what's supposed to happen. For example, "Our customers need to load their bike on the rack" could be translated into a symptom, with an outcome and impact:

> "[Our customers] may [not be able to fit their bike to the rack], which leads to [frustration]."

Translating needs into symptoms may be an obvious exercise, but it's still useful because we break down symptoms even more later which helps us with design inputs and priorities. We don't want to collect just the easy symptoms. Take some time to consider what our user is trying to accomplish. What possible unintended outcome is going to lead to an unpleasant experience? At the same time, we need to stay within the scope of the concept space.

Other sources of symptoms may be complaints. Analyze complaints about previous versions of your product or competitors' products. What is it that users complain about? They are directly reporting the problems they experience when using this product. Ask, "After using our product to achieve their goal, what are potential things users may complain about?" If it sounds like a complaint, you're probably working with symptoms.

Failures specific to the product (as yet undeveloped) are too soon to list.

We cannot really list failures of a concept design because we don't have one yet. At concept development, our product is so fuzzy that it's difficult to identify functions and failures. However, after we've collected ideas about design inputs by using the concept space, we'll likely have much more clarity about our concept design.

Different activities like failure mode and effects analyses (FMEAs) focus on product functions and potential failures. After our co-work in the concept space, we can begin those types of potential failure analyses with much more confidence that we will get meaningful results. (We will push our work of the concept space into FMEA in chapter 12.) After concept development, a better understanding of potential failures and their causes and effects will lead to additional design inputs as we develop the product.

Hazards are top-down negative events beyond the scope of the Concept Space.

Risk evaluators usually categorize hazards as things like biological hazards, physical hazards, and safety hazards. Your industry may use different categories specific to your product type and use environment.

Hazard analyses focus on what could go wrong and include sources of hazards and their causes. They are used to identify risk controls which may include some design choices.

Identifying the sources of hazards is beyond the scope of the Concept Space. In the concept space, we develop a concept design focused on the user's experience with our product. We are not looking to identify hazards. Rather, we are looking for potential experiences that make our users unhappy.

Collect ideas and avoid circling down the path of hazards.

In our example, our customer got frustrated. What if they got hurt, like a cut? Could the cut, if left untreated, lead to an infection, sepsis, and potential death? That is a potential outcome, although rare. But we are not performing a hazard analysis. We are gathering design inputs against which to design, so we don't have to go so far along the causal chain of events.

We can instead focus on the symptoms most likely to be experienced by our customers. This not only keeps our ideas close to the concept space, but it also keeps our customer's experiences top of mind.

Plus, designing against symptoms at the point our customer experiences them will also prevent more serious outcomes. We'll be stopping—or at least slowing—the chain of events at the source by creating design inputs that control against those risks. When in the concept space, focus on the more immediate symptoms our customers may experience when things go wrong.

Team Activity: Gather a list of symptom statements.

Look for information about complaints to get started. If you don't already have a list of symptoms, consider developing them now according to the following steps. Your goal at the end of the working session is to have several symptom statements.

These steps combine the ADEPT Team Framework and Concept Space Model over one ADEPT cycle:

Do pre-work. Gather customer descriptions and preliminary needs. Review the ADEPT Team Framework to get ready to lead the meeting.

1. **Align** on the scope and goal of the working meeting. Make the goal visible on agendas and during teamwork. An example: "A list of [number] potential symptoms our customers experience when our product does not perform as we expect."

2. **Discover** potential symptom statements.

   [Our customers] may [lose product features, be challenged], which leads to them [experiencing this negative impact].

   Brain write ideas and then share them. Use affinity diagrams to group and refine ideas so you can examine them in the next step.

3. **Examine** the list of symptom statements. A facilitator summarizes the group's findings for understanding and clarity.

4. **Prioritize** the symptoms for continued exploration of design inputs.

   4a. Multi-voting: Display the list of symptoms. Each team member gets five dots (sticker or drawn) to place on their choice of symptoms to explore more for design inputs. They should not put more than one dot on any one symptom.

   4b. Review and discuss the results. Aim for consensus, which is that place where everyone supports the decision even if it wasn't their first choice. If there is no consensus, consider going back to the Examine step to ensure all team members are voting on the same idea or the same understanding of an idea.

5. Perform the **Teamwork** activities to close the meeting. List actions for follow-up, including items noted on the side in the "parking lot."

6. Summarize this team design activity in writing and share what you've learned.

   6a. Summarize symptom statements and their priorities and share with the team.

   6b. Update user descriptions and their needs.

Once you've completed this ADEPT cycle, you'll have a list of symptom statements. You'll also have team agreement on which symptoms to continue exploring for next steps.

Use process endpoints, steps, and decisions are not the same thing.

The way our customers use our product is the Use Process of our design. They follow steps to accomplish tasks. There is a lot we can understand about our concept ideas just by looking at high-level use steps. I'm not referring to the things the product itself does, but the basic steps our user takes to get from start to finish.

A flowchart is a visual map of process steps and decisions. It's also known as a process flow diagram and is one of the seven basic quality tools. Many manufacturing, industrial, and business operations use flowcharts.

I use flowcharts frequently on my own and with teams. When built out with a team, it provides a lot of clarity and prompts a lot of discussions. Plus, there are many ways we can analyze the flowchart to derive design inputs depending on what it is we want to learn.

Construct a basic flowchart.

To create a flowchart, there are some basic symbols you need to know to get started. The first is probably the smallest: the **arrows.** They connect shapes and show what happens in sequence.

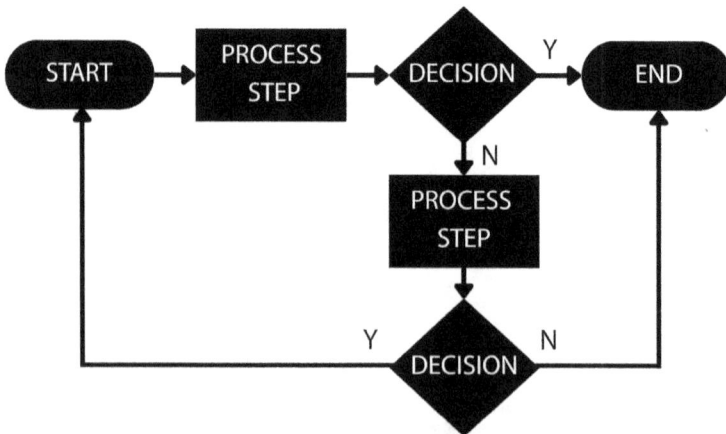

Figure 7.2. Basic Flowchart Shapes

The start and an end of a flowchart are designated as **ovals**. They define the scope of what it is we're analyzing. They're the bookends of the process. We always need to have one start and one end.

Process steps are **rectangles**, which represent functions. We usually label them with a verb and a noun, like "lift this" or "assemble that." Finally, decisions are represented by a **diamond**. A minimum of two arrows should attach to a decision to represent the two options related to decisions—for instance, Y for Yes and N for No. We usually label them near the relevant arrow.

Flowcharts are a great team exercise. They're very accessible and visual, so people can understand what it is we're trying to accomplish rather quickly. That also makes it easier to invite people to contribute, arrange things, and prompt discussions about the use process.

We can have an in-person meeting or meet virtually with online white boards to create flowcharts with our team. Whether real or virtual, I recommend using Post-it notes instead of flowchart software. This encourages teams to move ideas around. Later, for posterity or record keeping, we can make the flowchart fancy with a software tool.

Analyze the flowchart for design inputs.

The start and end points are something we need to define to keep ourselves focused on the user, not the product. We're asking, "What is it that *the user* is doing?" We're not asking, "What is it that *our product* is doing?"

We can derive a lot of design inputs from this type of flowchart at concept development before we even really know what we're designing. Building out a flowchart with our cross-functional team can ease concept development by helping us:

- **Align:** A flowchart helps our team agree on a scope and use steps. Then we all understand what our customers expect to gain out of using our product.

- **Get clarity about users:** We may have one particular end user, but through the flowchart, we may realize that there is somebody else that's interacting with our product during the use process. A flowchart can help us identify users, each of whom may have different needs.

- **Find performance gaps:** A flowchart can help us understand differences and gaps. We can explore what is ideal versus actual, what our product does versus our competitor's, or how we want to upgrade our version 2.0 product.

- **Identify interface requirements:** What else does our concept product need for the job to get done? A flowchart can help us identify the steps and when and where they need to occur.

- **Prioritize:** We can prioritize what needs to happen for a successful outcome versus what we ask users to do. We can also look at costs and quality as part of our use process to optimize our product.

- **Find knowledge gaps:** A flowchart gives the team ideas of what to ask users. When we start this flowchart with our team, we may end up with more questions than we have answers for. But that's a good thing because we're learning more about the use concept space so we can design for it.

The whole reason for the process flowchart is to get alignment with our team and then set expectations for what our product can do.

Team Activity: Draw the potential use process.

You may already have a potential use process you want to explore. You may also have a marketing analysis or customer surveys that were done when the project was first approved.

If you don't yet have a use process, consider hosting a separate team working session to develop it with the following steps. Your goal at the end of the working session is to have mapped out a high-level process flowchart that includes a clear beginning and end, process steps, and decisions. Plus, your team will prioritize what they want to learn for design, which will guide you to which flowchart analysis to do next.

Follow the steps to complete an ADEPT cycle for the use process. These steps use one ADEPT cycle:

Do pre-work. Gather customer descriptions and preliminary needs. Review the basic flowchart shapes. Review the ADEPT Team Framework to get ready to lead the meeting.

1. **Align** on a scope and goal of the working meeting.

   1a. Make the goal visible on agendas and during teamwork. An example: "Create a high-level flowchart of the use process for this new product."

   1b. Agree on a scope. Define the beginning and end of the process. Refer to the Concept Space to help define the scope of the use process. Clarify that the team includes ideas of process steps and user decisions, even at the concept level of design.

2. **Discover** potential use process flow.

    2a. Each team member brain writes ideas of process steps and decisions. What are activities the user needs to do with the product to get from start to finish?

    2b. Team members share their ideas, putting them in use order.

    2c. Use affinity diagrams to group and refine ideas. Group ideas into process and decision steps.

3. **Examine** the use steps. A facilitator summarizes the group's findings for understanding and clarity.

4. **Prioritize** the next flowchart analysis by gaining consensus on what the team wants to better learn for design right now.

    4a. Alignment: Has the team identified new needs or users, without knowing how it affects the use process? If so, dive deeper into that issue with an alignment exercise.

    4b. Comparison: Does the team want to make improvements from an existing process, like your potential product versus a competitor, version 1.0 versus 2.0, or ideal versus actual? Prioritize reaching consensus on the use process.

    4c. Critical to Quality: Does the team want to list important functional priorities and interface requirements that are critical to product quality? If so, prioritize a Critical to Quality analysis.

    4d. Value-Added: Does the team want to list value-added functions for development and list non-value-added functions to downgrade, eliminate, or make the use process easier? Prioritize a Value-Added Analysis.

4e. Deployment: Does the team want to separate functions by user group, list interface, transfer, or coordination requirements? Prioritize a Deployment analysis.

4f. Review and discuss the results. Aim for consensus, which is that place where everyone supports the decision even if it wasn't their first choice. If there is no consensus, consider going back to the Examine step to ensure all team members are voting on the same idea or the same understanding of an idea.

5. Perform the **Teamwork** activities to close the meeting. List actions for follow-up, including items noted on the side in the "parking lot."

6. Design for the use process. Summarize this team design activity in writing and share what was learned. Use teamwork results to link needs with requirements.

6a. Summarize the use process flowchart and share with the team.

6b. Update user descriptions and their needs.

Once you've completed this ADEPT cycle, you'll understand basic steps your user takes to get from start to finish using your black-box product. You'll also have team agreement on which analysis to perform for next steps.

Keep the use process high-level for the Concept Space Model.

Our team is developing the next generation bike stand, and we're using process flowcharts to explore the concept space for design inputs.

The first thing we decided are start and end points for the use process. The starting point is when the bike stand is delivered to the doorstep, and our endpoint is when the bike stand is disposed of.

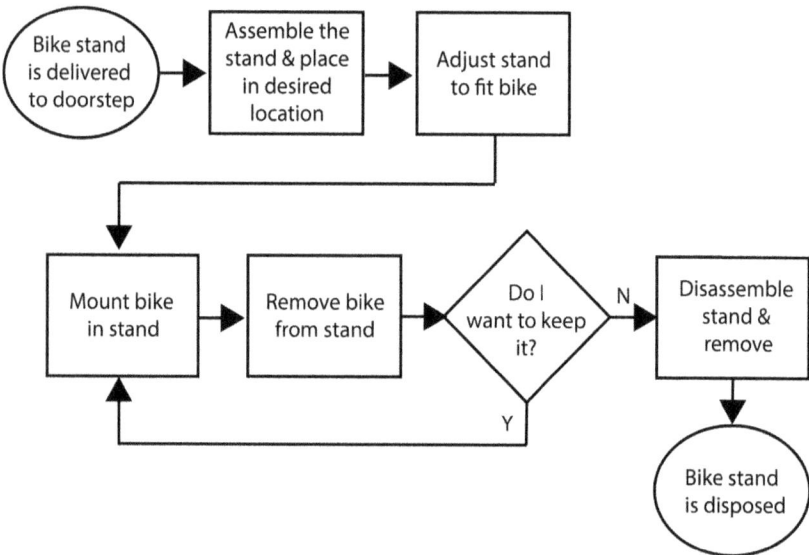

Figure 7.3. The high-level use process for our Example Scenario

We then diagram several high-level use steps by using our rectangles and diamonds. We assemble the stand, adjust it, mount the bike in the stand, and remove it from the stand. Eventually our users are going to think, "Do I really want to keep this?" We've captured that as a decision step. If they don't want to keep it, they need to disassemble the stand and return or otherwise dispose of it.

We use this flowchart to ensure the team is aligned on the scope of the use process—the beginning and end points—and the high-level process steps associated with our users completing the tasks to be done.

ADEPT Team Framework cycles do not require separate meetings.

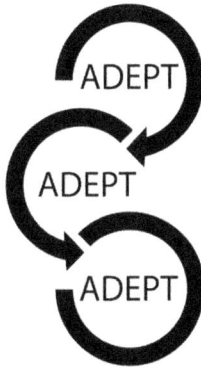

Figure 6.8.

I broke down the previous steps into three separate ADEPT cycles. I usually think of each ADEPT cycle as its own meeting. However, especially for this Concept Space Model, these cycles can take place in one meeting.

How many meetings it takes depends on your project scope and your team. If the scope is clear and your team is synergized, then keep the team together and work through the benefits, symptoms, and use process in one sitting. Think of your meeting as having sections, with each section being an ADEPT cycle.

It's also possible to break it up into two meetings. In the first, focus on benefits and symptoms. In the second, focus on the use process. This really depends on your situation.

You're leading this effort, so you can decide. It may be that you thought you could do it all in one meeting, but it's not going well or taking way longer than you think. You have permission to change your mind and ask your team to regroup to finish the Concept Space Model activity. Break it off at the end of an ADEPT cycle and make plans with your team to meet another time.

You can also ask your team how to best work together. If they say it fits their schedules better to have short bursts of co-working sessions, then plan for one cycle per meeting. If they'd rather team up for an extended period, then plan to combine multiple ADEPT cycles.

If you do combine multiple ADEPT cycles into one meeting, just watch for team exhaustion. Doing these activities is a lot of work, and your teammates may need a break!

You should have distinct ideas at the end of an ADEPT cycle using the Concept Space Model.

Here is an example of a team exercise to identify potential benefits, symptoms, and a basic use process for a bicycle stand. You can see it's a combination whiteboard and Post-it note activity.

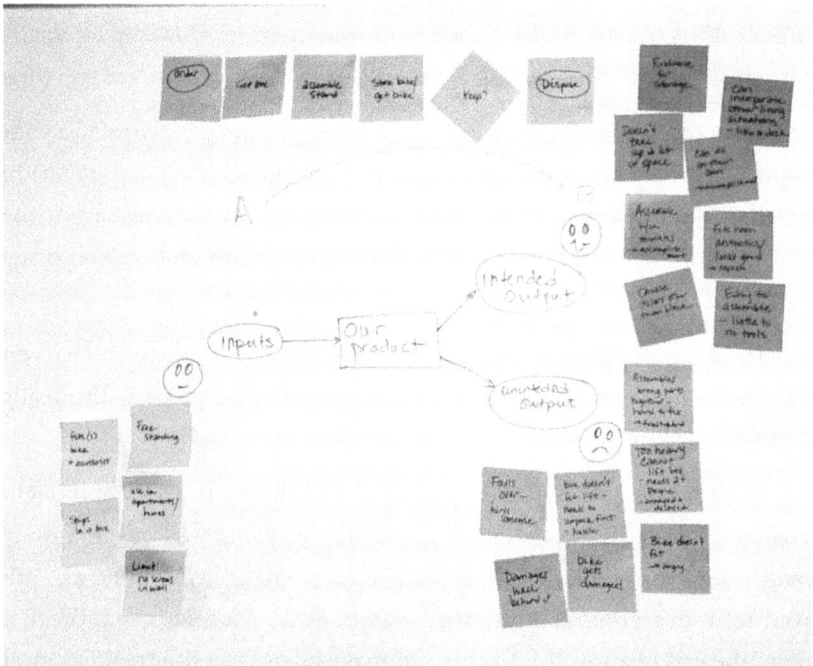

Figure 7.3. How the Concept Space Model may be laid
out on a whiteboard with Post-it notes.

Key Takeaways:

1. The Concept Space Model is used to get alignment on users and the use environment by understanding where a new product will be used and what it will achieve.

2. Use the ADEPT Team Framework to help plan and execute co-work sessions. Apply an ADEPT cycle for each Focus.

3. Needs, benefits, and features are different, although related. Focus on benefits with the Concept Space Model, which will be targeted customer experiences. The team can prioritize which ones to explore more.

4. Symptoms, failures, and hazards are different, although related. Focus on symptoms with the Concept Space Model, which will be customer experiences we target to prevent or reduce. The team can prioritize which ones to explore more in the next step.

5. We don't need to know features to map out a high-level use process. Capture ideas now for further development. The team can prioritize what further analysis would most help add clarity and understanding.

6. Do not spend too much time trying to fully understand the customer first, as a better understanding of the customer will emerge from discussions of the other three goals.

Reflection Questions:

1. How well do I understand the users, their assumptions, and the use environment for my current project, and what specific areas do I need to explore further with my team?

_____

_____

2. Can I clearly differentiate between the needs, benefits, and features of my product, and how does my team's understanding align with mine?

_____

_____

3. How well do I differentiate between symptoms, failures, and hazards? How does my team's understanding compare with mine?

_____

_____

4. How well do I understand the processes our customers go through when using our products?

_____

_____

Chapter Eight

DISCOVER CUSTOMER DESIRES AND DESIGN FEATURES THEY'LL LOVE

Design doesn't need to be delightful for it to work, but that's like saying food doesn't need to be tasty to keep us alive.

— Frank Chimero, author of "The Shape of Design" and thought leader on the craft of creation

—⊰◯∭◯⊱—

IN THIS CHAPTER, I'll tell you how to use the targeted Benefit-Impact Model and Template with your cross-functional team. We start by breaking apart the benefits into features and their impacts. Then, we prioritize potential features against a level of customer satisfaction. We further prioritize a level of design implementation using a Kano Model.

We have intended outputs of our product, including a list of benefits. But what does that mean to our users? What kind of impact would that make on them?

We are going to break apart our targeted benefits into features and their impacts. This exercise helps divide out our thinking processes and stay focused on what it is we want to learn. Breaking them down into their parts helps us and our team *really* get into some interesting things that can help form our design inputs for *great* designs.

Break down benefits.

Benefits are a step between our customer needs and the technical design inputs. We're going to break apart user benefits so we can get drivers not only for the product design itself, but what would also affect other parts of the product offering. If we develop our product from benefits, we're keeping the user in mind in both how they'll use our product and the positive outcomes they'll experience because they used our product.

Evaluate what creates benefits to gather design inputs.

Let's review how to use this Benefits-Impact Model with our teams for design inputs. We reorganize our benefits into an equation, describing benefits as a product feature that gets used and the impact that feature's use has on our user. We can associate the impact with a customer satisfaction rating. With our team, we identify the features and their impact.

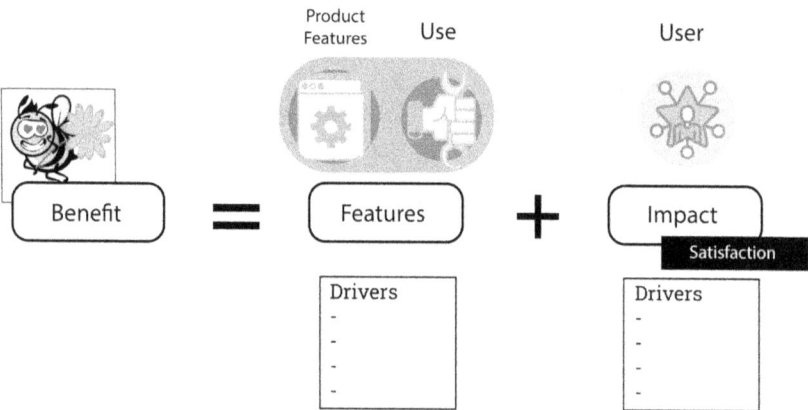

Figure 8.1. The Benefit-Impact Model with Drivers and Priority

Next, we explore drivers as a step between customer needs and design inputs.

When examining drivers to benefits, we are essentially exploring the reasons why a particular benefit can be achieved and how its positive impact on the customer can be amplified. Think of it this way:

Features are the "what" of your product—the specific capabilities or characteristics it possesses.

Feature drivers are the "how" behind those features. They are the underlying mechanisms, design choices, or supporting elements that make a feature possible. When we identify feature drivers, we start to generate concrete ideas about what our product needs to include or do to deliver that feature.

For example, if the benefit is "easily moveable," feature drivers might be lightweight materials, incorporating wheels or casters, or modular design. These drivers then lead to potential design inputs like materials, including wheels, or modularity.

Drivers to benefits also include impact drivers. These drivers are opportunities to enhance the impact and customer satisfaction. Drivers make the impact greater. These are not always direct product features but can include related services, packaging, or even understanding the use environment.

To continue our example, we decide the key impact for our customers is to "have a sense of control over their space" after easily moving our product to where they want. An impact driver may be that users can rearrange their environment on their own, without assistance. This is something we can add to our concept design.

In simpler terms, when we consider the drivers of a benefit, we're asking:

- What needs to be true about our product's design or related offerings for this beneficial feature to exist?

- Assuming this feature is present, what other factors can contribute to a greater positive impact for our users?

By answering these questions, we move beyond just stating the desired benefit and begin to generate actionable ideas for specific product features and other opportunities to make that benefit a reality and maximize its positive impact. These ideas then serve as potential design inputs.

To develop benefits, we now have a Benefit-Impact Template.

This template expands the Benefit-Impact Model by adding boxes for where we fill in information. First, we define our Benefit Statement by filling in the feature and impact. Then, we list the drivers below that. Finally, we relate the impact to a customer satisfaction rating.

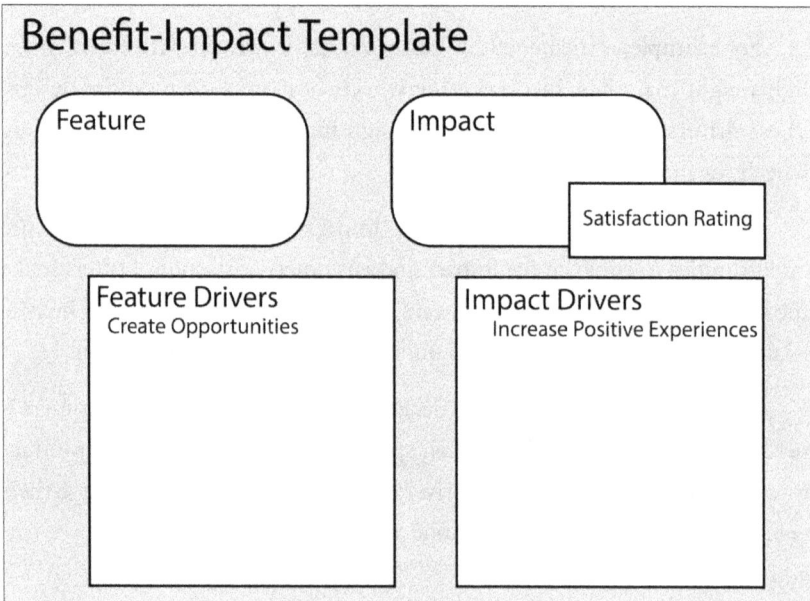

Benefit-Impact Template

Feature

Impact

Satisfaction Rating

Feature Drivers
Create Opportunities

Impact Drivers
Increase Positive Experiences

Figure 8.2. Benefit-Impact Template

Don't worry too much about which drivers go where. The main reason for using this template is to get our team to think about the product, its use, and also the customer's experience with it, given that our product has a particular feature. It breaks down a sizeable chunk of information to come up with different ideas. Ultimately, we take the results and start developing our design inputs.

Example Scenario: Define the feature and impact and list drivers.

The best way to show the value of breaking apart benefits for design inputs is with an example.

Figure 8.3. A Bicycle Stand Design Scenario

Our scenario is that we're part of a team that is developing the next generation bike stand. Right now, our model is to ship the bike stand in boxes for the customer to assemble. We've heard statements about customer experiences we want to target, statements like "Customers want this to be an easy assembly."

Where do we come up with ideas with our team? I'm sure you've already come up with ideas of your own, but remember we want to engage our team in developing design inputs. We'll also learn more about our customers along the way.

We first identify and gather our team. When we meet, we align on the scope and goal of evaluating a potential benefit related to this feedback. Our goal is to complete a benefits statement and develop design inputs. We draw or write out the benefits equation or the skeleton of the benefits statement.

How can we turn this statement into a benefits statement? "Customers want this to be an easy assembly." What fact-based features are we really evaluating, and what sort of impact is this going to have on our customers?

We decide "Customers can assemble the bike stand within minutes of receiving the box" is our target design feature. It is fact-based, measurable, and describes how the user is interacting with our concept bike stand. We decide this as a team based on the market research, surveys, and experiences we've had with our customers.

What would be the impact on our customers, given that they have this feature and can assemble the stand within minutes?

With team input, we understand our customer wants to use the bike stand quickly with an accomplished feeling. So, our Benefit Statement is "Our customers can assemble the bike stand within minutes of receiving the box so they can use the product quickly, gain confidence in their decision, and feel a sense of accomplishment."

Now, we brain write ideas for drivers for the feature and the impact.

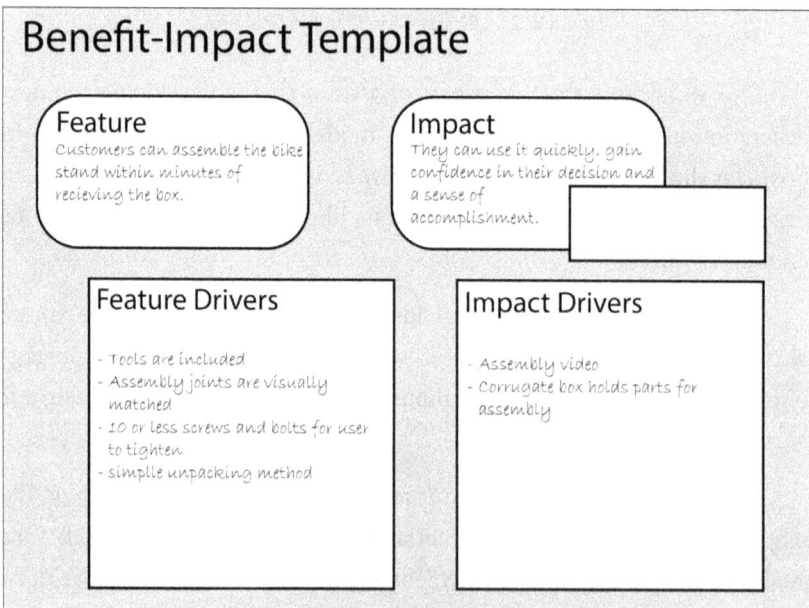

Benefit-Impact Template

Feature
Customers can assemble the bike stand within minutes of recieving the box.

Impact
They can use it quickly, gain confidence in their decision and a sense of accomplishment.

Feature Drivers
- Tools are included
- Assembly joints are visually matched
- 10 or less screws and bolts for user to tighten
- simplle unpacking method

Impact Drivers
- Assembly video
- Corrugate box holds parts for assembly

Figure 8.4. The Benefit-Impact Template for our Example Scenario

What design choices of our bike stand would affect the feature "Customers can assemble the bike stand within minutes of receiving the box"? Our team has identified including tools in the packaging, assembly joints being visually matched for easy identification, limiting the number of screws or bolts, and using a simple unpacking method.

Given that this feature is available, what can enhance our user's impact—or what about our offering can increase the likelihood the customer experience is that "They can use the product quickly, gain confidence in their decision, and feel a sense of accomplishment"? Our team has identified an assembly video and a shipping box that helps hold the parts during assembly.

With these drivers, we've identified design inputs against which we can engineer a product. We've also identified other ways we can enhance the customer experience.

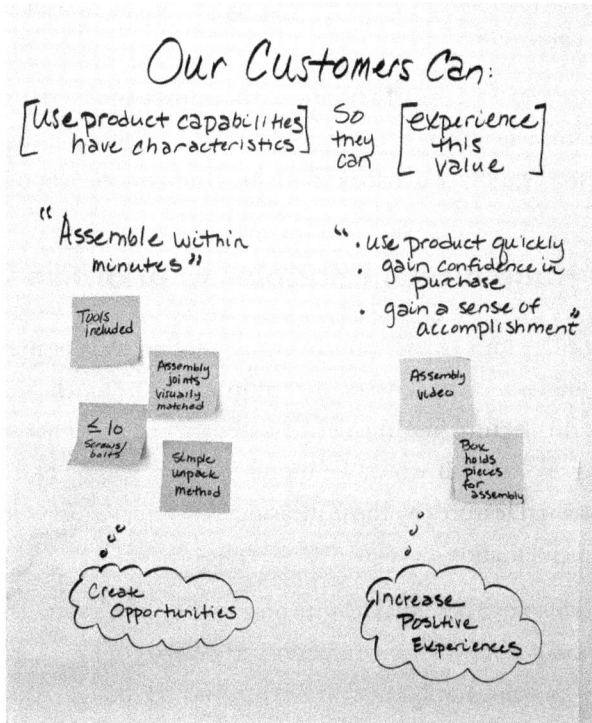

Figure 8.5. A layout example of the Benefit-Impact Template on a Whiteboard with Post-its

Are there other benefit statements that could apply to "Customers want this to be an easy assembly"? If so, we can develop more benefits statements around this same feedback. With a few benefit statements developed, we could revisit our customers with surveys, focus groups, or other conversations to validate that we're developing the right product with the benefits that they would like.

Prioritize design inputs using Benefits-Impact.

The Benefits Model also helps us prioritize what really matters for this product. This is because we can prioritize our features based on the impact they provide using a satisfaction rating. This is not unlike other team exercises where we break out an effect of an unintended output into its parts to analyze risk and prioritize risk factors. But from that point of view, we control bad things from happening through our design choices. With a benefit breakdown, we encourage good things to happen through our design choices.

There are two ways we can prioritize the impact and its related features with our team. One way is to assign a customer satisfaction rating. Another way is to use a two-by-two matrix. Both are based on the Kano Model.

Use the Kano Model Principles to prioritize impact.

To prioritize impact, we want to create a rating scale or mark priority on a two-by-two chart. We're determining how much we want to implement this feature into the design based on the customer satisfaction it will bring. For this, it would be useful to use a two-by-two chart that assesses potential features by those measures: customer satisfaction and the level of implementation.

We can map our potential feature on a two-by-two chart. The y-axis of the chart would be customer satisfaction. The x-axis of the chart would be our level of implementation of that feature into the design. These are also the axis of a Kano Model.

A Kano Model typically displays arrows illustrating how product characteristics and their implementation can affect customer satisfaction. The overarching idea of the Kano Model is that not all design inputs are equal.

Not all customer statements have the same weight, and not all the characteristics we decide to develop should be implemented equally. The farther along the positive x-axis, the more effort will be invested into ensuring the design meets this benefit. The higher the mark in the upper-right quadrant, the more priority the design input should take over others.

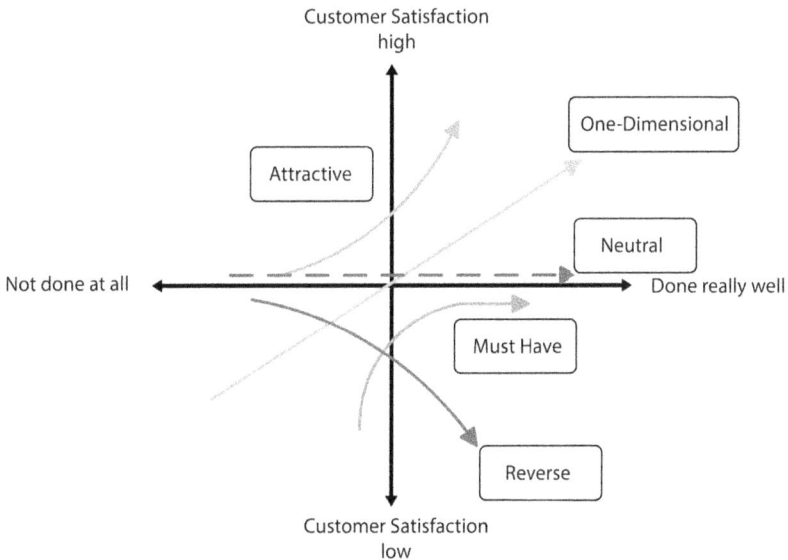

Figure 8.6. The Kano Model

A common example of a Kano Model in practice is a feature index. When you are a consumer and evaluating whether to buy a product, you may come across a feature index. The feature index shows different options you could buy into. Those options are usually features.

Table 8.1. Feature Index Example of an Offering
Embodies the Kano Model ideas

Base Model	Super Model	Deluxe Model
X	X	X
X	X	X
	X	X
		X
		X

With the base model, you'll likely have the features you need to do the job (must-have features). However, choosing one of the other options has additional benefits. The middle option likely has features that provide benefits you are really interested in (one-dimensional features). The top option likely has the features that would dazzle you (attractive features).

In concept development, we're not trying to create a marketing feature index. We want to work with our team to target certain customer benefits. We can use a Kano Model matrix to help a team assign a level of customer satisfaction associated with a feature's availability.

Doing this after we've listed drivers to those features and impact helps us to decide where to put the mark on the two-by-two matrix. What would really increase customer satisfaction? How can our product differentiate itself from the other competitive products in the field? How is it possible to excite customers and non-customers?

And on the opposite side of that, are there features or ideas we've come up with that we can back-off of implementing fully? What features can we eliminate because they're not providing customer satisfaction?

Let's survey the arrow lines on the Kano Model:

One dimensional

First, let's consider the middle, or the one-dimensional area, of our model. A line that goes from negative to positive through the center of the axes represents this. It's nearly a direct relationship between customer satisfaction and how well something is really done. These one-dimensional factors are really "wants." They lead to customer satisfaction when they're met and dissatisfaction when they're not met. In fact, the better they're implemented, the higher the customer satisfaction.

Attractive

Another impact measure, attractive, is above our one-dimensional line. The Attractive line starts in the "not done at all" area of the Kano Model. Then it quickly escalates to a high customer satisfaction with little movement into "done very well." If attractive factors are present in our product, they cause a high satisfaction with very little implementation. Attractive factors don't cause customer dissatisfaction. If we don't have these kinds of characteristics in our product, we may not differentiate our products from our competitors.

Must-Haves

Almost mirrored, but opposite, of the attractive line on the model is our must-have factors. Our must-have factors extend from a low customer satisfaction area into something that's neutral. If they're not included, they create customer dissatisfaction. When we include them, if they're done well they lead to something that's more neutral than satisfied. These factors don't increase the customer satisfaction, but our customers expect it. It's a must-have to keep those factors.

Neutral Factors

The neutral factors are indifferent. Our customers don't really care if they're there or not. They make little sense to implement because they don't influence customer satisfaction, or they may make it worse. But there might be a reason they're there.

Reverse Factors

We want to avoid reverse factors. Reverse factors are really a rejection of our product design. They dissatisfy the customer. And in fact, the more we have of it, the more dissatisfied they are. It can lead to a bad brand image.

Neutral and reverse factors are not adding value to customer satisfaction or are subtracting from it.

The Kano Model is a great visual model and template to use in concept development because it's going to help you further define your product characteristic priorities: things you want to manage in product management and product development, the decisions you make about quality controls later, and your internal customers.

Time shifts the results of the Kano Model.

The Kano Model is a snapshot in time. Time influences whether a characteristic stays in one area of the graph or it shifts to another. An attractive, new feature can shift to a one-dimensional or must-have line in the Kano Model. That's because what was novel and new, over time, becomes the norm. An obvious example is the evolution of the cell phone from flip phones to screens to flip screens.

Assign a customer satisfaction rating to a benefit-impact to prioritize.

You can use the principles of the Kano Model to create a rating scale. A satisfaction rating can help us with design priorities. We're not assigning a satisfaction rating number to do any mathematics with it. Rather, it is an easier way for us to prioritize our features and design inputs. We're going to apply that satisfaction rating to the benefit's impact to prioritize these features.

To more easily apply a satisfaction rating to our Benefits Model, we correlate each rating with a voice of the customer statement. For example, for a satisfaction rating of 5, the highest satisfaction, we could think that our customers would say, "I'm delighted! I didn't even know I wanted this or that this was possible!" Teams can immediately prioritize benefits based on a satisfaction rating alone. The higher the satisfaction rating, the more important it is to the design.

Table 8.2. A Customer Satisfaction Rating Scale

Satisfaction Rating	Impact "Voice of the Customer"	Outcome can be described as	Most likely applies to these measures of quality
5	Attractive "I am delighted or excited. I didn't even know I wanted this or that this was possible!"	The Outcome does not need to be fully implemented to delight the customer. If the Outcome is missing or insufficient, the customer doesn't notice.	• Aesthetics • Performance • Features • Empathy (e.g. caring and individual attention) • Responsiveness (e.g. willingness, prompt service)
4	One-Dimensional "I want this. The more of it or the better I get it, the more satisfied I am."	The Outcome has a direct relationship to Customer Satisfaction.	• Aesthetics • Performance • Features • Durability • Empathy • Responsiveness • Tangibles (e.g. service forms)
3	Must Have "I expect this. This must be, as a minimum. If it's done badly, I tolerate it, but I'm dissatisfied."	The Outcome can only satisfy the customer. As it becomes more available or of a higher quality, satisfaction increases but only to a neutral state.	• Conformance • Durability • Serviceability • Reliability • Assurance (e.g. service trust & confidence)
2	Neutral "This isn't important to me, but I don't mind having it."	The outcome doesn't improve or detract from customer satisfaction.	n/a
1	Reverse "I dislike this. It's frustrating. I would not consider a product with this."	The outcome leaves the customer dissatisfied or frustrated.	n/a

You may notice that our number 1 in our satisfaction rating is "reverse" and associated with something our customers don't like. When you happen upon these when you're exploring benefits, put it in the "parking lot" as out of the scope of that co-working activity. You want to come back to them later when you're using the Symptom-Impact Model. We include it in our satisfaction rating because the Kano Model uses it.

Alternatively, prioritize by plotting the benefit-impact on an empty two-by-two chart.

A Kano Model is a special two-by-two chart we can use to further prioritize benefits. Our team may decide to use a two-by-two chart to prioritize instead of a satisfaction rating. People use two-by-two charts to prioritize lots of different ideas. I'm sure you've heard, or have used, charts that evaluate urgent versus important, time versus impact, or value versus effort. These are all ways we can evaluate ideas and concepts with two different criteria.

A Kano Model relates customer satisfaction with the level of implementation. You'll ask your team to place a mark on the empty two-by-two chart associated with each benefit-impact. We are asking them to prioritize concept feature ideas. They identify the level of availability of that feature against the level of customer satisfaction it would provide. We then bring the co-working session to an end and analyze where they placed the mark to help us with design inputs.

Example Scenario: Assign a design implementation priority.

Let's revisit our scenario with the bike stand. Our benefits statement was:

> **Customers can assemble the bike stand within minutes of receiving the box so they can use the product quickly, gain confidence in their decision and a sense of accomplishment.**

After our team considers the impact and the drivers, they vote that the impact is one-dimensional and rate it a 4.

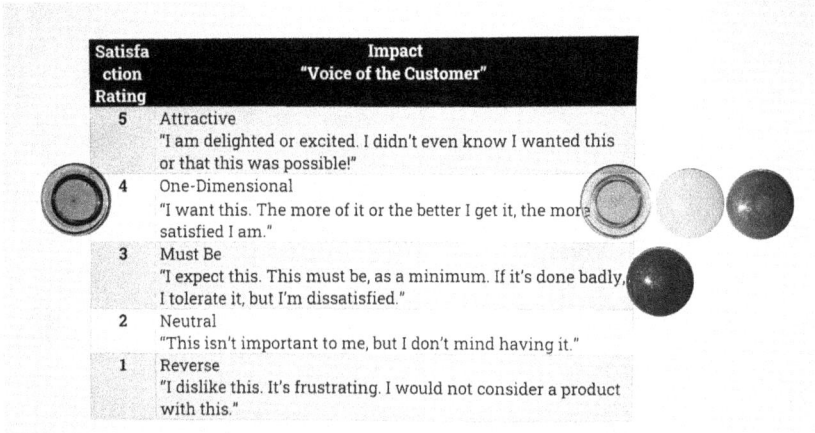

Satisfaction Rating	Impact "Voice of the Customer"
5	Attractive "I am delighted or excited. I didn't even know I wanted this or that this was possible!"
4	One-Dimensional "I want this. The more of it or the better I get it, the more satisfied I am."
3	Must Be "I expect this. This must be, as a minimum. If it's done badly, I tolerate it, but I'm dissatisfied."
2	Neutral "This isn't important to me, but I don't mind having it."
1	Reverse "I dislike this. It's frustrating. I would not consider a product with this."

Figure 8.7. Multi-Voting on a customer satisfaction rating for a Benefit-Impact

We can also use a two-by-two matrix, the Kano Model. Our team puts marks in the upper right quadrant of the matrix. The visual regression line of our team's choices leads us to conclude "one-dimensional" for the rating.

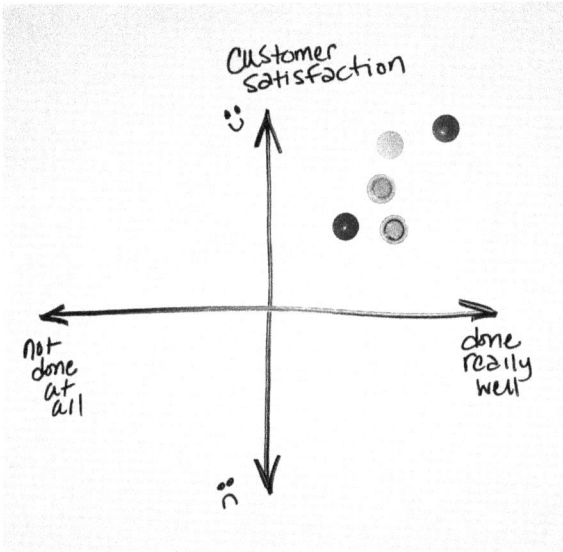

Figure 8.8. Multi-voting on a customer satisfaction rating with a 2x2 chart

Our Benefit-Impact Template is now complete.

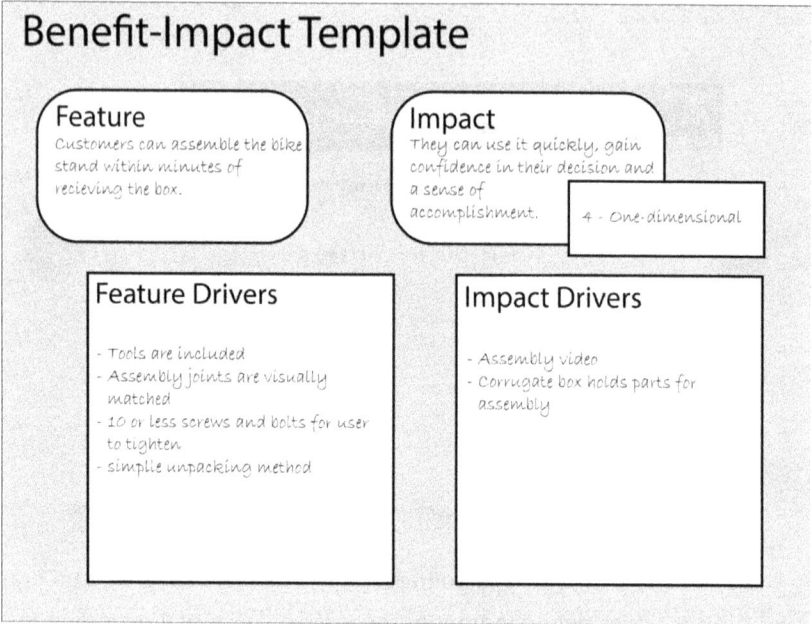

Benefit-Impact Template

Feature
Customers can assemble the bike stand within minutes of recieving the box.

Impact
They can use it quickly, gain confidence in their decision and a sense of accomplishment.

4 - One-dimensional

Feature Drivers

- Tools are included
- Assembly joints are visually matched
- 10 or less screws and bolts for user to tighten
- simple unpacking method

Impact Drivers

- Assembly video
- Corrugate box holds parts for assembly

Figure 8.9 The Benefit-Impact Template from our Running Example

Team Activity: Move from Benefits to Design Input Priorities.

With the Benefit-Impact Template and the Kano Model, we can move from benefits to design inputs and then evaluate the importance of those inputs to customer satisfaction.

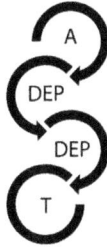

Follow these steps to gather design inputs for each benefit. These steps use an ADEPT cycle that include multiple Discover-Examine-Prioritize cycles for each item.

Do pre-work. Gather customer descriptions and preliminary needs. Review customer satisfaction rating scales. Review the ADEPT Team Framework to get ready to lead the meeting.

1. **Align** on a goal and scope for the working meeting.

   1a. Get team consensus on a qualitative rating of customer satisfaction using a Satisfaction Rating Scale, which should include a Kano Model category and the voice of the customer. Review the Kano Chart.

   1b. Make the goal visible on agendas and during teamwork. An example: "A list of [number] potential benefits our customers experience when our product performs as we expect; described with potential features, services, and use environment facts that may drive them; and prioritized by customer satisfaction."

1c. Agree on a scope. Which benefits are you exploring right now? Break out benefits into their features and impacts, using a benefits statement to help. Draw the Benefits Model or write the benefits statement(s) so the entire team can see it. Ensure you have highlighted each feature and impact.

2. **Discover** design ideas through drivers. Choose one benefit to explore.

2a. Each team member brain writes their ideas of how the product provides the capabilities and characteristics, or the drivers for each feature. How might we meet the capabilities? What characteristics of a concept design could help provide this feature? Give the team between five and 10 minutes to complete this step. Use Post-it sized notes to record the ideas.

2b. Next, each team member brain writes their ideas about the product or offering that can affect the impact or the drivers for each impact. What else could make this impact more likely or stronger? Give the team between five and 10 minutes to complete this step. Record the ideas on Post-it sized notes.

2c. Next, all team members share their ideas, placing them under the outcome and impact for all to read. If you're using a whiteboard, place them in the "drivers" area of the template.

3. **Examine** the drivers area of your template. To provide clarity, a facilitator presents the group's findings by summarizing the ideas for each outcome and impact. This also expands the group's understanding of the benefit statement.

4. **Prioritize** features. Relate the benefit-impact to a level of customer satisfaction.

    4a. Try Voting: Share printed copies of the customer satisfaction rating scale and ask team members to write down or mark the value corresponding to this benefit-impact.

    4b. Alternatively, draw a two-by-two chart with one axis labeled "customer satisfaction" and another "level of implementation." Each team member gets one dot (sticker or drawn) to place on the Kano Model, corresponding to this benefit-impact.

    4c. Review and discuss the results. Aim for consensus, which is that place where everyone supports the decision even if it wasn't their first choice. If there is no consensus, consider going back to the Examine step to ensure all team members are voting on the same idea or the same understanding of an idea.

5. Repeat the Discover, Examine, and Prioritize steps for other benefit statements.

6. Perform the **Teamwork** activities to close the meeting. List follow-up actions, including those related to the "parking lot." Take pictures or save whiteboards.

7. **Analyze the results of the co-work sessions.** The written report is a record of your activities and shows how their co-work in evaluating benefits is fitting into design decisions. List and prioritize features and impacts. List the potential feature drivers and offerings and draw a Kano Model. Keep the feedback loop going and maintain an open mind, viewing this as another opportunity to follow up and follow through.

These steps cycle through Discover, Examine, and Prioritize for each benefit. At the end, you'll have a list of benefit features with a priority based on customer impact. You'll also have ideas of features and offerings related to that benefit. We now have:

- Benefit statements, each of which include a feature and its benefit for the customer.

- Drivers, which are the underlying mechanisms, design choices, or supporting elements that make a feature possible.

- More drivers or ideas that provide opportunities to enhance the impact of a feature on customer satisfaction.

- Each impact with a rating that establishes a relationship between benefit and customer satisfaction, This also allows for a priority ranking to be compared with other benefits.

Now, we propose design inputs.

Make choices about design inputs based on the Benefits-Impact Model co-working session.

With all this information gathered with our cross-functional team, we can really explore design benefits and their drivers and start drafting design inputs. We can continue to use this information throughout the development process. Remember that we use this co-work activity with our cross-functional team as an opportunity to learn before we tackle engineering the design.

Take these next steps to ensure you incorporate your team's contributions into design decisions:

1. **Update user descriptions for the project along with their corresponding needs.** Your team may have uncovered new things during the exercise, so be sure you capture that in your user descriptions. This co-work session may also lead us to more questions that we might want to go back and ask our customers.

2. **Share your insights with your team.** Hold a discussion where you discuss your recommendations about the level of implementation. In the meeting, show how your team's co-work in evaluating benefits fits into design decisions. Keep the feedback loop going along with an open mind, viewing this as another chance to learn what your teammates know.

3. **From Feature Drivers, develop design concept options.**

    - What about the design concept could you create or change to make this a feature of the product?

    - Are there any repeated drivers among the benefits? Focus on these ideas to maximize your design effort. This approach can lead to multiple benefits with just one feature.

4. **From Impact Drivers, develop concept offerings.** Offerings may exist outside of the design and benefit other functions of the business.

    - What offerings of the product or service could increase positive customer experiences and its impact?

    - Design can impact the offerings. Check how your design can simplify things. Be cautious of how your design may complicate things.

5. **Develop design inputs**. Maintain the information to iterate throughout the development process.

Use the Kano Model to develop design inputs.

There are many ways you can develop information from Benefits into concept design decisions and design inputs. The Kano Model in particular can provide helpful information at this stage of your process.

Set or adjust the level of planned implementation based on your team's impact assignment.

If your team used a numerical rating scale to assign a rating to the impact, then prioritize design inputs based on that scale. If your team used a two-by-two chart, then closely examine the mark's placement on the Kano Model to determine which category it fits in. Now, consider what that effect has on design decisions.

If we have a characteristic we think is attractive, we could think of that as critical to motivating our users to buy and use our product.

For characteristics that are one-dimensional, we could think of them as critical to satisfaction because the more we have of those characteristics, the more satisfied our customers are going to be.

If we have a "must-have" factor, we can think of that as critical to quality because if we don't have that, our customers will not think our product is of sufficient quality.

Neutral factors are not critical, but they might be necessary—and we definitely want to avoid reverse factors.

Table 8.3. Design decisions relating to a Customer
Satisfaction Scale and the Kano Model

If the impact is...	Consider doing this for engineering implementation:
Attractive	The design inputs that correspond with this benefit are Critical to Motivation (CTM). This feature can differentiate your design from the rest of the market. Explore the drivers to the feature and impact to discover ways to implement and enhance this. Over time, these features will become expected and drift to Must Have.
One-dimensional	The design inputs that correspond with this benefit are Critical to Satisfaction (CTS). You want the design to have these features. The more or better the features are implemented, the more satisfied the customer will be. How can you best implement this feature in your design? What are the features of similar, successful products?
Must Be	The design inputs that correspond with this benefit are Critical to Quality (CTQ). This must be an outcome of the design, but it may not need to be fully implemented or the best. Discuss the level of implementation needed to satisfy the customer. Where along the Kano Model's x-axis would the Must-Have line level-off? Weigh efforts and costs to make it is fully functional against the level needed to satisfy the customer.
Neutral	Consider reducing or eliminating the implementation of these features. Are they an added cost with no benefit? Do we really need this? The primary user may not care, but it may be needed for the business, regulations, or to function. Ensure your team considers all the customers before removing it.
Reverse	Actively reduce or eliminate these features to avoid the impact. Look at drivers to the features as things to avoid. What kind of design measures may lessen their impact?

Compare the impact of design benefits to set design direction.

An assigned rating using the Kano Model is just one input into project management and setting design priorities. Understanding the priority of each benefit can help us prioritize and make a list. We can also compare benefits against each other when we plot them on a Kano Model.

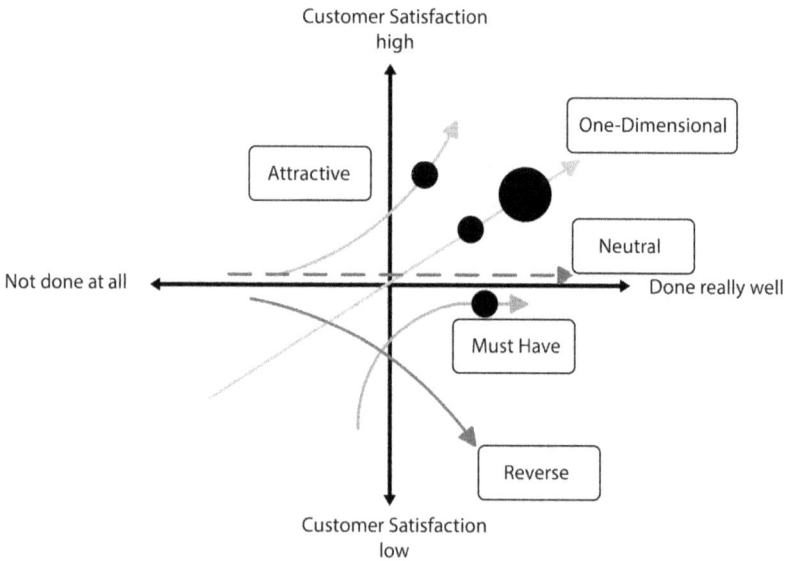

Figure 8.10. Using a Kano Model to map Benefits and prioritize for Design Implementation

If we don't have any attractive benefits, then we may need to do more research because those are critical when it comes to motivating people to buy and use our product. We may need to identify more characteristics that have appeal.

If we have many one-dimensional factors, then we may need to continue researching to identify which one takes priority. We have limited time, resources, and money to develop this new product. Is there a characteristic that takes priority over the others? Do we need to get more information from our customers?

For the must-have characteristics, what steps can we take to ensure that they are implemented? We can also think of these characteristics as critical to quality, which is going to affect the development of our product. We may need to monitor and control these features for quality to ensure they are included in product development at the level our customers need.

For neutral characteristics, which maybe the customer doesn't care about, are they needed for the product to be manufactured? We may need to consider if internal parties value these aspects of the product, even if the customer does not.

Keep in touch with your team.

Maintain communication about your design decisions with your team as often as you can. Being able to link your design to their input helps ensure the optimal design for the user, and it eases the buy-in needed once the design is completed. It also helps you make design decisions about other things like materials, placement, and testing.

Key Takeaways:

1. A benefit statement links products and users and provides more specific information about the targeted user experience.

2. Benefits can be broken down into features and their impacts.

3. Feature drivers are the underlying mechanisms, design choices, or supporting elements that make a feature possible. Impact drivers include opportunities to enhance the impact of a feature on customer satisfaction. These are not always direct product features but can include related services, packaging, or even understanding the use environment.

4. Prioritize design inputs by using a customer satisfaction rating or a two-by-two matrix based on the Kano Model.

5. The Kano Model is a snapshot in time, since an attractive feature can shift to a one-dimensional feature or "must-have" feature over time.

Reflection Questions:

1. How do I understand customer benefits? Can I link a potential benefit to a customer impact?

   _____

   _____

2. What customer satisfaction scale will I use to help me prioritize benefits based on the impact? Do I have a scale, or do I need to create a new scale? (Consider adapting the ones in this book.)

   _____

   _____

Chapter Nine

UNCOVER POTENTIAL PROBLEMS AND DESIGN PRODUCTS THAT EXCEED EXPECTATIONS

Having true wisdom means preventing difficult problems from arising in the first place.

—Thomas Huynh on Sun Tzu's Art of War

———————

IN THIS CHAPTER, I demonstrate how to use the Symptoms Model/ Template with your team. Symptoms will be described as an outcome and its impact. We will identify various drivers for symptoms, leading to different actions for risk prevention or loss reduction. Finally, we will assess symptom risks, gather information, and prioritize design actions.

In early concept development, we do not have specific failures or causes to examine because we don't have a product yet. However, we have potential symptoms, or negative customer experiences, that could happen when things don't go right.

There are many things we can bring to light by identifying and looking at the potential symptoms. Breaking down symptoms into their parts helps us and our team get into some interesting insights that can help form design inputs.

Break down symptoms.

Let's talk about how we can describe symptoms as an outcome and its impact. This is going to help us better understand situations and evaluate them for risk. Reorganizing our symptoms statement into an equation helps us see its parts. We can describe symptoms as an undesired outcome and the impact that has on our user. We can associate the impact with a severity rating. The outcome must happen first for there to be an impact.

[Our customers] may [lose product features, be challenged], which leads to them [experiencing this negative impact].

Let's consider a previous example: "Our customers may not fit their bike to the rack without taking off the bike's accessories, which leads to frustrationand regret with their purchase." In this example, the outcome is "not fit their bike into the rack without taking off the bike's accessories." The impact is "frustration."

Evaluate what drives potential symptoms to get design inputs.

Examining symptoms and their drivers also helps us identify ways to block risks and reduce losses. It gives us a chance to explore the underlying reasons and circumstances that could lead to a negative customer experience and what might influence the severity of that negative experience.

Outcome drivers are the "how it happened." They're the factual reasons, potential failures, or usage scenarios that could cause the undesired outcome to occur. When we identify outcome drivers, we start to generate concrete ideas about which aspects of our product's concept might lead to these problems.

For example, if an outcome is "Our customers cannot move the product into different rooms," drivers could be bulky dimensions or complex disassembly/reassembly required. These drivers point towards the need for design features that avoid bulky dimensions or complex assemblies.

Impact drivers are product or service offerings that can make the impact worse. When we identify impact drivers, we create ideas about how we can lessen the symptom's impact.

For instance, in the same example, the impact is "frustration and regret with purchase." Impact drivers could include unmet expectations and damage to their walls. These drive the impact to be greater. The more our customers' expectations aren't met and the more damage that occurs, the more they are frustrated and regret the purchase. It also suggests a potentially medium-high severity, as the customer's opinion of the product (and potentially the brand) is negatively affected.

Impact drivers suggest opportunities to reduce the negative impact despite a bad outcome. In our example, we may devise ways to communicate better with the customer and ensure they understand the product's limitations before they buy it.

In simpler terms, when we consider the drivers of a symptom, we're asking:

- What could realistically cause this specific negative event (the outcome) to happen with our product concept?

- If this negative event does occur, what factors could make the resulting negative experience (the impact) worse for our users?

By answering these questions, we move beyond just identifying potential negative experiences and begin to generate actionable ideas for understanding and addressing the root causes of those outcomes and mitigating their impact. These ideas then serve as potential design inputs aimed at preventing risks and reducing losses. Outcome drivers help us think about how to block risks, while impact drivers help us develop prevention controls to reduce loss and impact.

To develop symptoms from design inputs, we have a Symptom-Impact Template.

This template expands the Symptom-Impact Model so we can add information. First, we define our Symptom Statement by filling in the outcome and impact. Then, we list the drivers. Finally, we relate the impact to a severity rating, which I will explain how to assign.

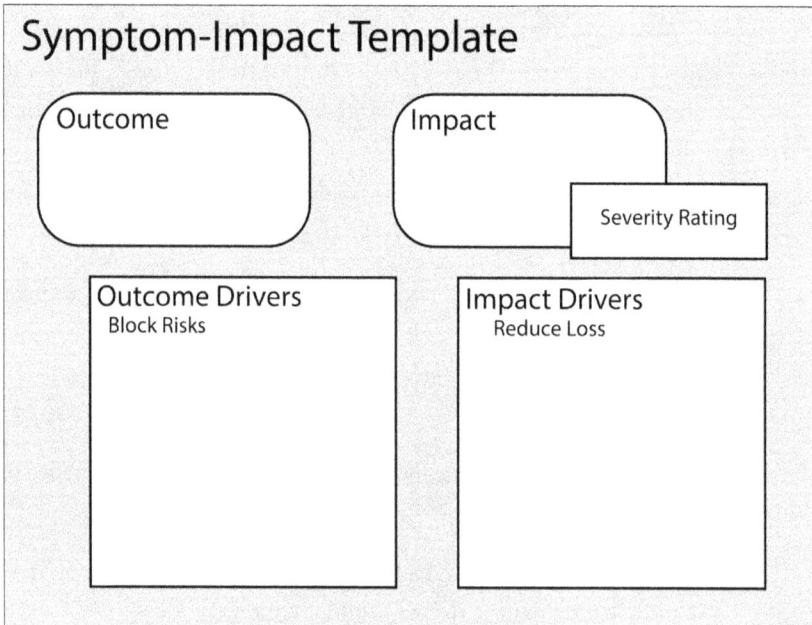

Figure 9.1. Symptom-Impact Template

Example Scenario: Discover and Examine Drivers for Design Inputs.

The best way to show the value of breaking apart symptoms for design inputs is with an example.

Figure 8.3

Our scenario is that we're part of a team developing the next generation bike stand. Right now, our model is to ship the bike stand in boxes for the customer to assemble. We've heard statements about customer experiences we want to further evaluate—for instance, "We've had many complaints that customers couldn't assemble the stand."

If we look closer at our symptom, we can break it into two parts: the outcome and its impact.

- Outcome: Customers cannot assemble the bike stand.

- Impact: Being unable to assemble the bike stand means they can't use it right away and need to spend their time resolving the issue.

Let's take a fresh look at the drivers so we can act. After all, this is what we want to do with these analyses. We want to get information to decide on what to do next or how much we need to do. In our example, we have our outcome and impact. When we think about drivers, we can think of them as asking this question: "What leads to the outcome?" and "What makes it more likely to happen?"

Potential drivers for this outcome are that the parts are wrong, damaged, or missing. Maybe the assembly instructions confuse customers. There could be too many parts in the packaging or a lack of tools needed for assembly. These are all factors that could cause a negative outcome.

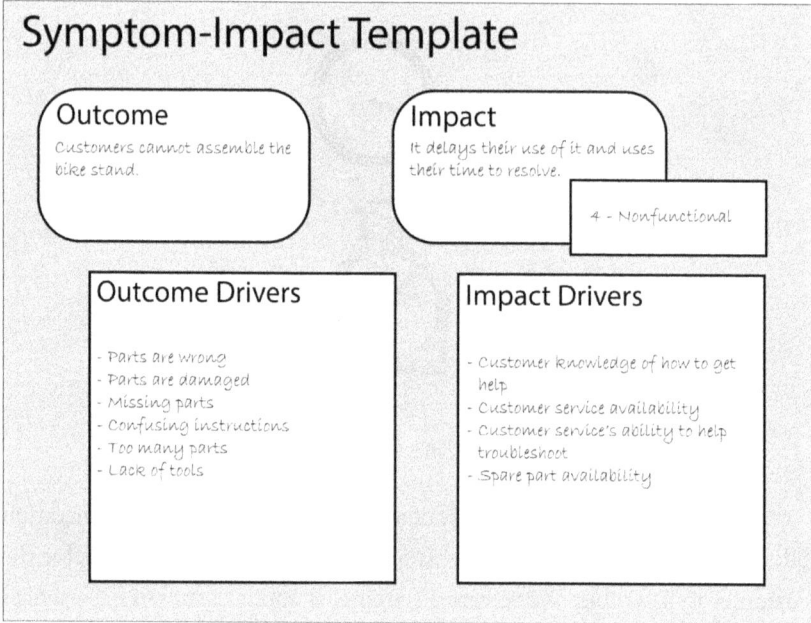

Symptom-Impact Template

Outcome
Customers cannot assemble the bike stand.

Impact
It delays their use of it and uses their time to resolve.

4 - Nonfunctional

Outcome Drivers
- Parts are wrong
- Parts are damaged
- Missing parts
- Confusing instructions
- Too many parts
- Lack of tools

Impact Drivers
- Customer knowledge of how to get help
- Customer service availability
- Customer service's ability to help troubleshoot
- Spare part availability

Figure 9.2. Symptom-Impact Template for our Example Scenario

Now, let's look at impact. A driver to this impact is our customers being able to get the help needed in a timely manner so they can finish the project. Other drivers could be the availability of customer service and their ability to help troubleshoot the problem and get to the root cause of customers' experiences.

Another driver for this impact could just be spare part availability in case customers can't assemble the bike stand because the wrong product was in the package or there was a damaged component. These are all drivers that can affect the likelihood and severity of the impact our customers experience.

In looking at these lists of drivers for outcome versus impact, we can identify the need for different activities to address each set of drivers. For outcome drivers, our goal is to mitigate risks and prevent negative outcomes. This proactive approach aims to tackle problems before they manifest. On the other hand, for impact drivers, our focus is on reducing losses and minimizing the effects of adverse events.

Assign a Severity to Prioritize an Outcome-Impact.

You can use the principles of risk management to create a rating scale. A severity rating can help us with design priorities. We're not assigning a severity rating number to do mathematics with it. We're going to apply the severity rating to the symptom's impact, which is an easier way for us to prioritize our design inputs.

The following scale uses dimensions of quality to relate customer and product quality to severity. Use a severity scale that applies to your industry, product, and project.

Table 9.1. A Severity Rating Scale Example

Rating	Customer	Product
6—Unsafe	Customer is injured. Customer does not trust/**is not assured** that the product is safe.	Product safety feature is not functional—**does not conform** to safety requirements.
5—Unreliable	Customer cannot use the product at all. Customer's opinion is that the product is **not reliable** or doesn't do what was promised.	Product is not functional and **cannot be repaired**. Is **not reliable in use**.
4—Nonfunctional	Customer cannot use the product until assembled or repaired. Customer relies on company **responsiveness** for assembly or repair.	Product is **not functional**. Product requires **repair**.
3—Underperforming	Customer needs to constantly adjust product for fit. **Empathy experience** is negatively affected—product design does not align with what they need.	Product has **reduced functionality**. Product is **not durable**. Product is **missing expected features**.
2 - Unappealing	Customer can use the product as-is with a defect but notices it. **Tangible experience** is negatively affected.	Product is fully functional but has a defect. **Aesthetics is negatively affected**.
1 - Neutral	No negative effects for Customer.	Product is functional with no defects.

Alternatively, prioritize by plotting the outcome-impact on an empty scatterplot.

To further prioritize symptoms, we can use a scatter plot. Instead of using a satisfaction rating, our team may choose to chart the severity and its likelihood to prioritize symptoms.

To establish a relationship between the severity of a symptom and its likelihood, ask your team to place a mark associated with each outcome-impact on an empty scatterplot. At the end of the co-working session, we analyze where everyone placed their marks to gather insights for design inputs.

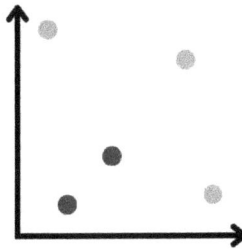

Figure 9.3. a scatterplot, of Severity and Occurrence as an example

Team Activity: Move from Symptoms to Prioritized Design Inputs.

By breaking out symptoms and relating a severity rating to the impact, we can prioritize information for design inputs.

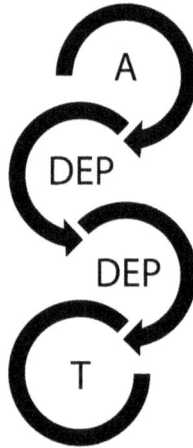

Follow these steps to gather design inputs for each symptom. These steps utilize an ADEPT cycle containing multiple Discover-Examine-Prioritize cycles for each item.

Do pre-work. Gather customer descriptions and preliminary needs. Review severity rating scales. Review the ADEPT Team Framework to get ready to lead the meeting.

1. **Align** on a goal and scope of the working meeting.

    1a. Get team consensus on a qualitative rating of severity of impact by using a Severity Rating Scale.

    1b. Make the goal visible on agendas and during teamwork. An example: "A list of (number) potential symptoms our customers experience when our product does not perform as we expect; described with potential outcomes, services, and use environment facts that may drive them; and prioritized by severity."

1c. Agree on a scope. Which symptoms are you exploring right now? Break out symptoms into their outcomes and impacts, using a symptoms statement to help. Draw the Symptoms Model or write the symptom statement(s) so that the entire team can see it. Ensure you have highlighted each outcome and impact (use brackets, underline, or otherwise highlight).

2. **Discover** design ideas through drivers. Choose one symptom to explore at a time.

2a. Each team member brain writes design ideas that can drive the outcome to be present. Limit the time allotted for this step to between five and 10 minutes. Restrict your idea-recording to Post-its notes or other small notes.

2b. Next, each team member brain writes the design ideas that can drive its impact to the customer. What other related factors can drive the impact or increase its chances of occurring? Limit the time allotted for this step to between five and 10 minutes. Restrict your idea-recording to Post-its or other small notes.

2c. Next, all team members share their ideas, placing them underneath the outcome and impact for all to read.

3. **Examine** the drivers to symptoms for clarity and understanding. A facilitator presents the group's findings by summarizing the ideas for each outcome and impact. This expands the group's understanding of the symptom statement and makes it clearer.

4. **Prioritize** symptoms. Relate the Symptom-Impact on a level of severity to the customer.

4a. Try voting: Share printed copies of the severity rating scale and ask the team to write down or mark the value corresponding to this benefit-impact.

4b. Or draw a chart with one axis labeled "severity" and another "likelihood." Each team member gets one dot (sticker or drawn) to place on the chart, corresponding to this symptom-impact.

4c. Review and discuss the results. Aim for consensus, which is that place where everyone supports the decision even if it wasn't their first choice. If there is no consensus, consider going back to the Examine step to ensure all team members vote on the same idea or the same understanding of an idea.

5. Repeat the Discover, Examine, and Prioritize steps for other symptom statements.

6. Perform the **Teamwork** activities to close the meeting. List follow-up actions, including those related to the "parking lot." Take pictures or save whiteboards.

7. **Analyze the results of the co-work sessions.** The written report is a record of your activities and shows how everyone's co-work in evaluating symptoms fits into design decisions. List and prioritize outcomes and impacts. List the potential drivers and offerings. Keep the feedback loop going, along with an open mind, viewing this as another opportunity to follow up and follow through.

These steps cycle through Discover, Examine, and Prioritize for each symptom. At the end, you'll have a list of features with a priority that's based on customer impact. You'll also have ideas of features and offerings related to that benefit.

Our entire goal of our co-working sessions is for knowledge sharing toward design inputs. We should have completed a session with our team. We now have:

- Symptom statements, each of which includes an outcome and its impact to the customer.

- Drivers (factual reasons, potential failures, or usage scenarios) that could cause the undesired outcome to occur. More drivers, or ideas, that can increase the impact, including factors affecting how customers experience negative outcomes.

- Priority of the impact associated with that outcome.

Now, we propose design inputs.

Make choices about design inputs from the Symptom-Impact Template co-working session.

With this information, our cross-functional team can really explore symptoms and their drivers and start drafting design inputs. We can also continue to use it throughout the development process. Remember we use this co-working activity as a learning opportunity with our cross-functional team before we do engineering design.

1. **Update the user descriptions for this project, along with their corresponding needs.** Your team may have uncovered new information during the exercise, so be sure you capture that in your user descriptions. This co-work session may also lead us to more questions that we want to ask our customers.

2. **Share your insights with your team.** Host a discussion in which you discuss your recommendations about the symptoms that are

important to control through design. In the meeting, which is a discussion not a formal co-working session, you show how your team's co-work in evaluating symptoms fits into design decisions. Keep the feedback loop going, along with an open mind, viewing this as another opportunity to learn what your teammates know.

3. **From Outcome Drivers, develop design concept options.**

   What about the design concept could you create or change to block risks?

   Are there repeats among the symptoms that you could focus on to maximize your design effort and the controls?

4. **From Impact Drivers, develop concept offerings.** Offerings may exist outside of the design and include other functions of the business. Are there any additional business functions that could help mitigate the impact of this situation on the product or service?

5. **Set or adjust the level of planned implementation based on your team's impact assignment.** When looking at potential problems in a design, the severity of these issues directly correlates with how bad they are. Therefore, we want to implement features in the design that can help control the impact of problems with the highest severity ratings. By addressing these high-severity issues proactively, we can mitigate the risks and ensure a more robust and reliable design. User satisfaction, trust, and product quality all improve with this approach. Focusing on high-impact features makes the design better and more robust.

6. **Develop design inputs**. Continue to use the information for iterations throughout the development process. Keep in mind that the most severe problems may be critical to product safety.

Keep in touch with your team.

Stay in communication with your team about your design decisions as often as you can. Being able to link your design to their input helps ensure a design for the user. And it facilitates the buy-in you need once the design is completed. Staying in touch with your team also helps you make design decisions about other things like risk controls and testing.

Key Takeaways:

1. Symptoms can be described as undesired outcomes and their impact on the user.

2. Outcome drivers are evidence-based reasons why an outcome could occur. Impact drivers are reasons why an impact is more likely to occur.

3. Use a severity rating or chart to prioritize symptoms, using principles of risk management.

Reflection Questions:

1. How do I understand current customer symptoms? Can I link a potential symptom to a customer impact?

   _____

   _____

2. What scale will I use to help me prioritize based on the impact? Do I have a scale already, or do I need to create a new scale? (Consider using your existing risk management methods or adapting the ones in this book.)

   _____

   _____

3. For a given symptom, can I distinguish between drivers that lead to the undesired outcome and those that affect the severity of its impact? How can I leverage this understanding to develop targeted design inputs?

   _____

   _____

Chapter Ten

MAP THE USER JOURNEY TO DESIGN SEAMLESS EXPERIENCES

"Design is not just about making things beautiful;
it's about making things work beautifully."

— *Roger Martin. Management Thinker, author of "Playing to Win"*
and "The Design of Business," and proponent of integrative thinking

———⟨≡≡≡⟩———

IN THIS CHAPTER, I show you how to analyze flowcharts to help you and your team understand more about the use process. We review when to iterate from high-level concepts to more detailed analysis. We also choose a way to analyze flowcharts for priorities to aid us in design as well as the basic use process based on what you want to learn.

The way our customers use our product is the use process of our design. They follow steps to accomplish tasks. Since we're in concept development, we're not focusing on what the product itself does. We are focusing on the way our users interact with our concept product. It's the high-level, general process they take to get from the input to the output of our Concept Space Model.

As part of our Concept Space Model activity in chapter 7, we've already drawn out a high-level use process. Now, we want to prioritize those steps so we can focus on what's needed for design inputs.

We approach flowchart analysis based on what we want to learn.

There are many ways we can analyze our use process from the Concept Space Model. First, we consider what is it we really want to learn.

If the process steps are confusing, then we may need to create another flowchart with a narrower scope. We can break out the confusing steps and do a sublevel process flowchart. Do what you need to do with your team to better understand your use process. I call this an **Alignment Flowchart**.

If we have a competitor product with elements we like or don't like, we use a **Comparison Analysis**. With this analysis, we're discovering the future state of our product. We're also gaining information about the gap between current and future states. This helps us decide what's important to implement.

We may want to learn about the interfaces and inputs that affect the use process. In this case, we start with the use process flowchart from the Concept Space Model. Then we can continue with a **Critical to Quality Analysis**. That helps us identify steps that are critical to outcomes and affected by inputs.

If we want to understand where we need to simplify things, we use a **Value-Added Analysis**. This also helps us identify steps that add value or don't.

And if we need to better understand who is doing what at certain points in the process, we can use a **Deployment Flowchart**.

We're exploring a concept. So, like other things in development, the more we learn, the more detailed we can get. We can focus on the priorities of the use process at concept development. As we figure out how it is we're going to meet the needs of our customers, we can further refine design inputs and do more detailed flowcharts.

From these flowchart analyses, we can learn if our design needs to be adapted for different users. We can also start to evaluate if it must operate with other products and determine if any modularity is needed for its use. Prompts may be needed for our user to properly use our product. We can also reduce errors and make our design as mistake-proof as possible.

We can continually use and iterate process flowcharts as needed with our team so we can better understand what we're developing. We don't have to do process flowcharts on everything. We can prioritize what's important and use them as a tool for learning. All these analyses are exercises you can do with a team using the ADEPT Team Framework.

Reduce confusion with an Alignment Flowchart exercise.

It's likely that the use process is not clear in concept development. That doesn't mean we throw up our hands and call it a day. Consider what is new, different, or unknown to you currently. Identify a few steps where it would benefit your team to dig a little deeper into the use process.

Example Scenario

Let's use an example to explore more how to use a flowchart to evaluate the use process and get further alignment. In our previous work with the Concept Space Model, we defined a high-level process flowchart, shown here:

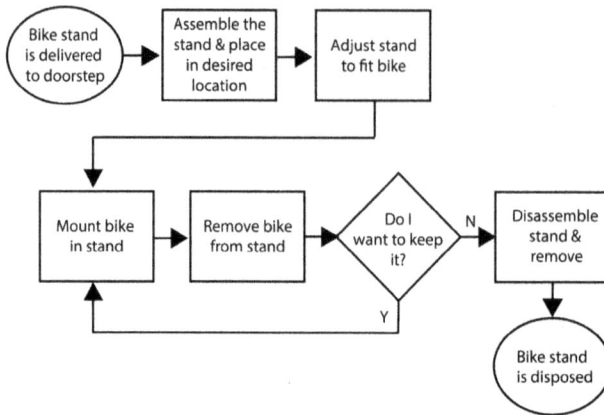

Figure 7.3.

After we created this high-level flowchart, we discover that we have some hidden customer needs. Our cross-functional team has been talking to our customers about this next generation bike stand. They're starting to get ideas about improvements our customers want.

Customers want to be able to adjust their bike stand while the bike tires rest on the floor. They don't want to lift their bicycles and then have to adjust the fit, which is what they do with other bike stands.

They also want a way to raise their bike in the stand so they can perform routine maintenance on the gears and wheels. They want the stand to make that easier to do.

These are two new customer needs that we want to evaluate. Can we design our next generation bike stand to meet these needs?

Since this is a brand-new customer benefit to our concept bicycle design, we don't really understand what it means to the use process. We look at our Concept Space Model use process to identify which steps these changes affect.

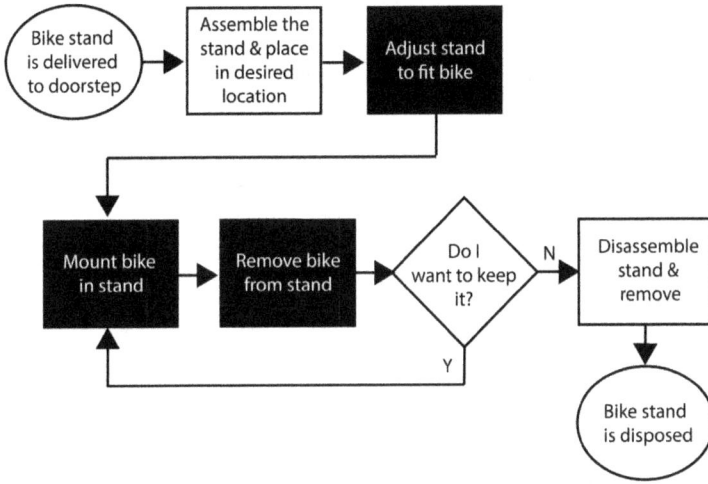

Figure 10.1. Identifying a New Focus to Explore for an Alignment Flowchart

We see these changes may not affect the bike stand assembly, but they do affect how the stand adjusts to fit the bike. Our customers want to adjust the stand with the bike on the floor, and they want to raise their bike stand so they can perform maintenance on it.

We decide we're really re-evaluating three steps: adjust the stand, mount the bike, and remove the bike. We're not sure how these new needs affect the use process. So, we gather our team and decide we want to focus on these three steps.

To re-examine our use process given this new information, we decide to create a new flowchart. First, we define new start and end points. Now, we're going to start at "the bike stand is assembled and placed," and we're going to end at "the bike stand is empty and it's still in its place in the consumer's house."

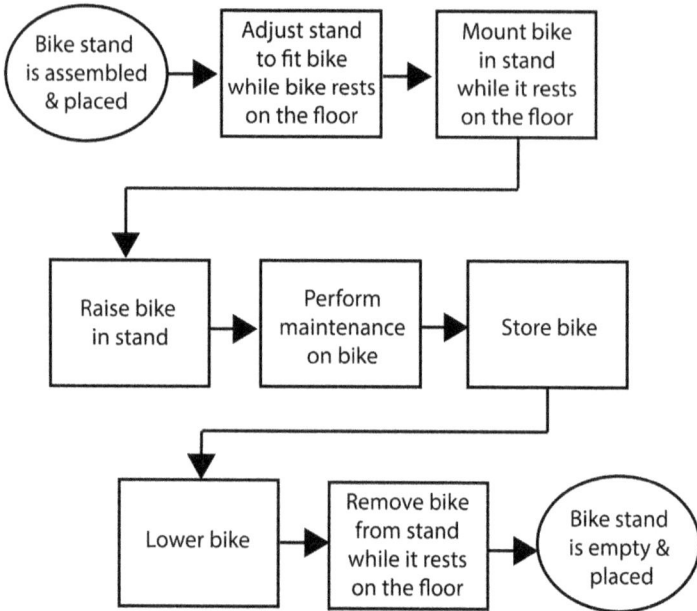

Figure 10.2. The Alignment Flowchart, showing areas to develop design inputs

Next, we list use process steps and decisions to explore what happens between these two points. We list possible steps in more detail, adding what the customer wants to do. We add "perform maintenance on the bike," "store bike," and "lower bike," and finally, "remove the bike from the stand while the bike rests on the floor." We add the condition "while the bike rests on the floor" to three steps.

Just these simple changes to the flowchart give us a better idea of which functions are affected by these new customer needs. Doing this is probably also going to raise a lot of questions, which we can research to better understand.

These are important design inputs for concept development. By working through this process with our team in a knowledge-sharing way, we achieve alignment with our team and what our customers want.

Team Activity: Analyze the use process with an Alignment Flowchart.

From your team's use process with the Concept Space Model, identify what's new, different, or unknown. Choose the related steps to focus on, then perform an Alignment Flowchart exercise with your team. At the end of this exercise, you'll have highlighted areas of the use process that need special consideration and focus during design.

Follow these steps to use the ADEPT Team Framework for an Alignment Flowchart exercise.

Do pre-work. Gather customer descriptions and preliminary needs, plus gather the flowcharts you'll use as the starting point. Review the basic flowchart shapes. Review the ADEPT Team Framework to get ready to lead the meeting.

1. **Align** on a goal and scope of the working meeting.

    1a. Review or make available the flowchart that will be analyzed.

    1b. Make the goal visible on agendas and during teamwork. An example: "Explore how new needs affect *these* process steps through drawing a new flowchart."

    1c. Agree on a scope. Describe the flowcharts you're reviewing. An example: "Draw a new flowchart which starts at [start point of flowchart] and ends at [end point of flowchart]."

2. **Discover** design ideas from the team.

   2a. Each team member brain writes ideas of process steps and decisions. What are activities the user would need to do with the product to get from start to finish, given this new scope?

   2b. Team members share their ideas, putting them in use order.

   2c. Use affinity diagrams to group and refine ideas. Group ideas into process and decision steps.

3. **Examine** the use steps. A facilitator summarizes the group's findings for understanding and clarity.

4. **Prioritize** changes. Relate the proposed changes for importance for design implementation.

   4a. Multi-voting: Display the process flowchart. Each team member gets five dots (sticker or drawn) to place on the flowchart, corresponding to what they believe is important to implement.

   4b. Review and discuss the results. Aim for consensus, which is that place where everyone supports the decision even if it wasn't their first choice. If there is no consensus, consider going back to the Examine step to ensure all team members are voting on the same idea or the same understanding of an idea.

5. Perform the **Teamwork** activities to close the meeting. List action items for follow-up, including "parking lot" items. Take pictures or save whiteboards.

6. Design for the use process. Summarize this team design activity in minutes and share what was learned. Use teamwork results to link needs with requirements.

   6a. Summarize important functional steps and use decisions and share with the team.

   6b. Update user descriptions and their needs, if necessary.

   6c. Prioritize implementation of use steps through features and offerings.

After the Alignment Flowchart exercise, you and your team will be clearer on how the new information about your users may affect the use process. With better understanding comes better design inputs.

Gather design inputs based on a comparison to an ideal with a Comparison Flowchart.

A Comparison Flowchart exercise helps your team compare your use process with another. Perhaps your team wants to make improvements, like your potential product versus a competitor, Version 1.0 compared to Version 2.0, or ideal versus actual.

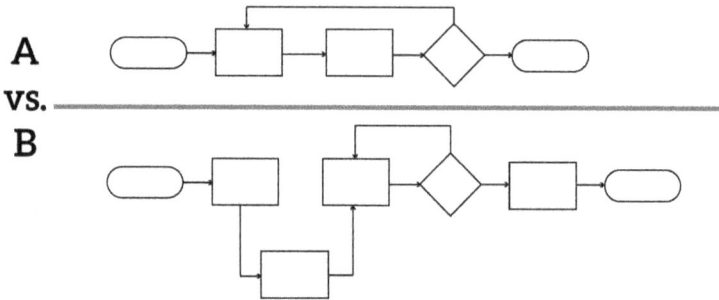

Figure 10.3. Comparison Flowcharts

To do this, draw two flowcharts: A and B. Or edit one flowchart and highlight the changes. Prioritize the use process by highlighting what is new, different, or unknown for investigation and development. Use multi-voting techniques to highlight priorities. Record activities for follow-up. By knowing the current state and mapping the desired future state, you gather information for design.

For the ADEPT Team Framework activity steps to perform a Comparison Flowchart Analysis, refer to Appendix B.

Identify important steps with the Critical to Quality analysis.

In a Critical to Quality analysis, we identify steps in our use process that affect our outputs. From this, we can identify which use steps are priorities for our users to achieve the output they want. We can focus efforts on making those process steps easy and robust.

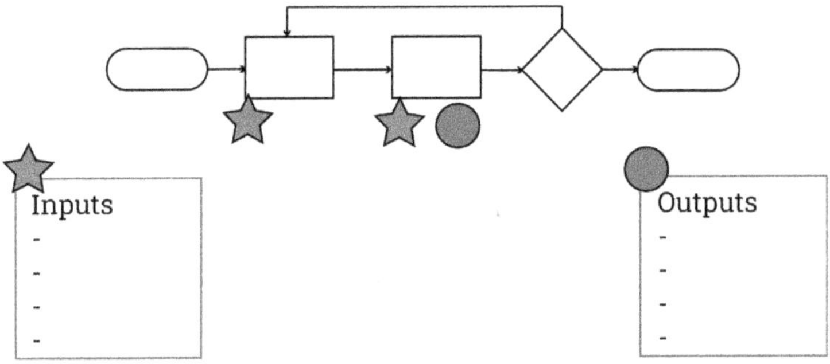

Figure 10.4. A Critical to Quality Analysis

We also evaluate the steps that are sensitive to the inputs, which helps us identify interface requirements for our concept product. It also gives us insight into sensitive steps, so we can make those steps more robust to changing conditions or find a way to better control those inputs.

By identifying what affects the outputs and what is affected by the inputs, we identify use steps that are critical to quality for our concept design. We analyze our flowchart for steps that are critical to quality and derive still more design inputs.

Example Scenario

Let's continue our bike stand example to demonstrate how to do this analysis. The scenario is that our team is developing the next generation bike stand, and we're using process flowcharts to explore the concept space for design inputs.

We've created alignment flowcharts and discovered we had new customer needs. We want to better understand some of the interface requirements, what affects the output, and what steps are affected by the input, so we'll do a Critical to Quality analysis on our flowchart.

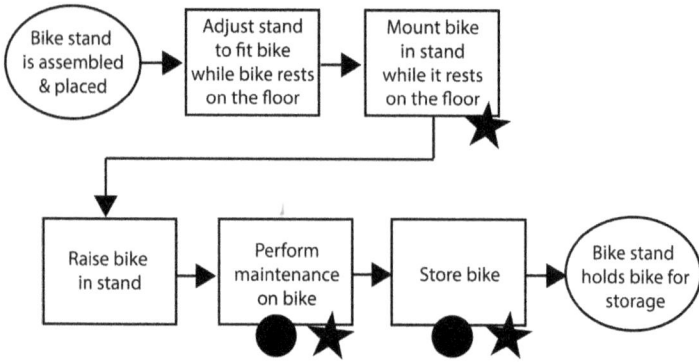

Figure 10.5. A Critical to Quality Analysis Example

Let's go through the analysis.

First, what are more specific inputs?

We meet customers where they are. Identify the who, what, when, where, and why at our process input. Doing this helps us clarify our users and further understand our use environment and other conditions and assumptions of our users.

When we look at inputs, the product itself is not an input. Instead, we look at things that are going to interface or interact with our product during the use process.

Using our example, here is what we consider:

- A bike is an input because it's separate from our bike stand, but it's going to interact with it. The status of our bike is that it's in need of repair or cleaning.

- Accessories on the bike are already properly adjusted by the user for riding. These include items like mirrors and baskets that are mounted on the bike and may interact with our bike stand.

- We note that our users have chosen a space in their dwelling where the bike stand will be stored.

- Since the bike needs repairs and routine maintenance, our users will also have tools, parts, and materials.

We can verify with our users and user groups that we have the right inputs.

Second, what are more specific outputs?

We identify the who, what, when, where, and why at our use process output. Doing this helps identify hidden customer needs or benefits.

In our example scenario, we have the following outputs:

- We still have our bike as an output that is stored and secured, which is one of the main reasons why our users want to have a bike stand. The user wants the bike to be in good shape and ready to use.

- The accessories are mounted on the bike and properly adjusted.

- At this point, our storage area is the same, with tools, parts, and materials stored in a tidy way.

To complete this step, we consider our users' goals. We anticipate how they would answer the questions, "What do you expect at the end of this process, when you're done doing these kinds of things? What are the things you expect to be able to do or to have happen?" This is how we think about our outputs. If we don't have ideas ourselves about how to approach the issues, we can always ask our users!

Third, what process steps affect the output?

This is going to help us identify the functional priorities of our concept design. Not all use steps are equally important to the output. Knowing what steps directly affect the output lets us identify the steps that need to be done well.

We want to find steps that hurt the output or help the output. In our example, based on the outputs provided by our customers, we think the step "perform maintenance on the bike" directly helps the outputs. The "storing the bike" function also directly affects the outputs. Mark those steps with a label or symbol. In our example, we use a dot.

Fourth, what steps are affected by inputs?

This gives us interface requirements. Someone, somewhere, is interacting with our product and affecting how it performs. We want to be able to identify those interfaces so we can design for them.

We want to identify the functional steps where the input affects what happens next in the process. In this case, we think "mount the bike in the stand while the bike rests on the floor" is affected by our inputs: "the bicycle," "the bicycle with all the accessories on it," and "the area of the dwelling where the person sets up the bike stand within their apartment." All those inputs affect that process step.

We also think "perform maintenance on bike" is a function that's affected by the inputs, because we not only have the bike, but now we also have the repair tools and the parts and the materials interacting with the bike. Consequently, with this functional step, our inputs are also going to affect "storing the bike."

Mark those steps with a label or symbol. In our example, we use a star.

Finally, study the results.

Now that our analysis is done, let's draw some conclusions.

Our process doesn't currently account for how to keep the storage area neat or where to store tools, parts and materials. How can we help the customer achieve this output? We can add a step to store the accessories and tools. That might be a new design input for us.

How can our stand help our customers perform maintenance? Is there something about the bike stand that could help with this part of our outputs? There may be some new design inputs regarding maintenance.

We have "accessories are on bike, properly adjusted." This is an input and an output, so we may be looking at a new customer need. Our newly identified need may be something like, "Customers need a way to store the bike with accessories so they don't need to adjust them when they want to ride it."

By identifying the process steps that affect the output, we prioritize the factors that are critical to quality. We also identify process steps affected by the input, which helps us better define the interface requirements. These are the kind of things we can do with a Critical to Quality analysis of a process flowchart.

(For the ADEPT Team Framework activity steps to perform a Critical to Quality Analysis, refer to Appendix C.)

Maximize customer satisfaction with a Value-Added Analysis.

The main purpose of doing a value-added analysis is to maximize our customer's satisfaction. We consider each step and highlight what adds value. Value-added steps are ones we definitely want to implement in our design.

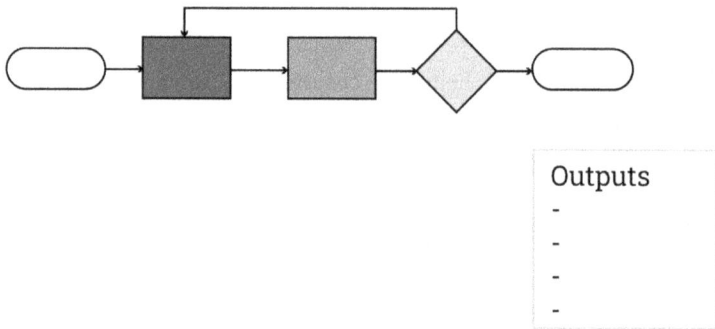

Figure 10.6. A Value-Added Analysis

In a Value-Added analysis, we categorize use steps based on the value they bring to achieving the targeted output: our customer's benefits. There are three levels of value that we identify for three different actions:

1. What adds value?

2. What does not add value?

3. What is necessary, regardless of whether it adds value or not?

Example Scenario

We continue our bike stand example to demonstrate how to do this analysis.

Our team is developing the next generation bike stand, and we're using process flowcharts to explore the concept space for design inputs.

We've already gotten some alignment by using an alignment flowchart, and now we want to look at what is value-added versus what is not value-added.

The value we're analyzing is associated with what contributes or detracts from customer satisfaction derived from our product. With a value-added analysis, we also get clearer on the outputs and the benefits, too, because it's going to prompt some discussions.

To do a value-added analysis, we first create our process flowchart within the desired scope.

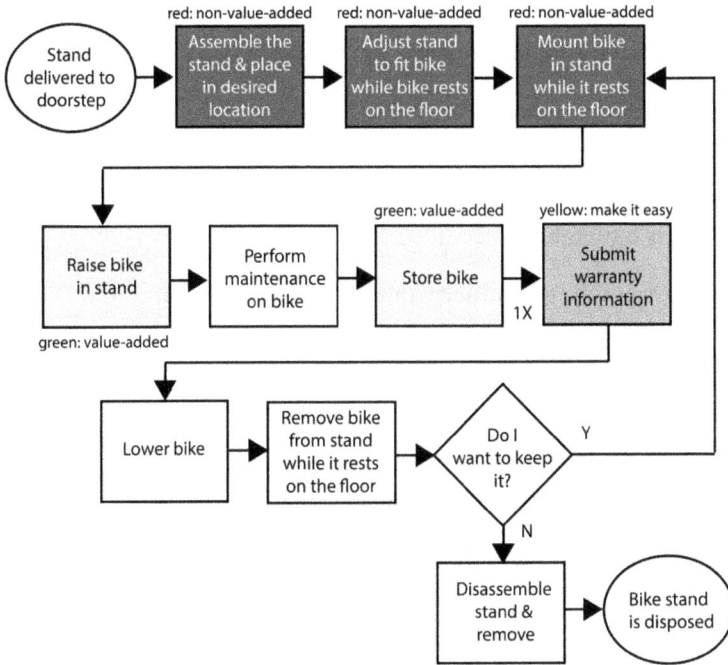

Figure 10.7. Value-Added Analysis Example

Next, we want to align on clear outputs. In this example, we use the same outputs that we developed for the Critical to Quality analysis. We identified the who, what, when, where, and why at our process output.

Then, we begin to analyze the use steps and decisions against the value they bring to achieving the output.

Consider: What adds value? Prioritize value-added steps to deliver benefits to customers in the best way possible.

These are high-priority steps that we want to ensure we implement and implement well. We identify steps that add value to the customer based on those outputs. We're going to highlight value-added steps. Most practitioners highlight value-added steps green.

In this example, we consider "raise the bike in the stand" and "store the bike." These are necessary steps to produce the kind of output our customers expect. These steps also contribute to customer satisfaction.

Consider: What does not add value? Reduce the effort or eliminate the step, if possible.

Next, we identify steps that do not add value. These are areas we want to simplify by reducing or eliminating them. We highlight these steps, too. In practice, most people highlight them red.

In our example, "assembling the stand," "adjusting the stand to fit the bike," and "mounting the bike" are all things that do not really add value to the output. They are things that need to be done, but they don't contribute to the kind of output the customer is looking for. These are the steps we want to be able to reduce or even eliminate, if we can.

Consider: What is an organizational need? Make these as easy as possible.

Finally, we want to identify steps that are necessary whether they add value or not. We want to reduce the obstacles to getting these steps done and make the process easier for the customer. We can highlight these in yellow.

For example, submitting warranty information is something we want to have happen, but the customer doesn't need to do this to use the bike stand. We can simplify this step.

Conducting a value-added analysis against the outputs of our process is something that can help us prioritize design concepts.

(For the ADEPT Team Framework activity steps to perform a Critical to Quality Analysis, refer to Appendix D.)

Deployment Process Flowchart Analysis

The final analysis of our process flowchart is a **deployment flowchart**. This happens when we have more than one user and we need to understand who is doing what. This can also help us understand and identify parallel activities. Mapping it out helps us identify design inputs.

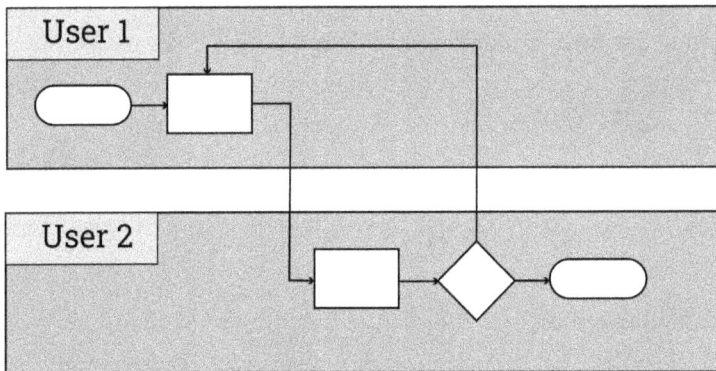

Figure 10.8. A Deployment Flowchart with two swim lanes

A Deployment Flowchart helps to answer important questions about our use process: Do we have multiple users? Do they work together to use our product to achieve something? Who is doing what and when?

We want our product to be used with success. Sometimes, that involves multiple users. An example is a physician using a medical device and a registered nurse prepping and assisting during the procedure. In a deployment flowchart analysis, we analyze the interaction between users to achieve success with our product during the use process. We also look at process steps from the point of view of the user who needs to perform them.

By shifting the use process flowchart steps into different swim lanes, we'll more clearly identify design inputs that are necessary because of different users. We'll be able to:

- Clarify users, use environment, customer needs, and interface requirements

- Coordinate use steps among multiple users

- Understand parallel activities

- Ensure proper transfer between steps

Example Scenario

We continue our bike stand example to demonstrate how to perform a deployment process flowchart analysis. As we've discussed, our team is developing the next generation bike stand, and we're using process flowcharts to explore the concept space for design inputs.

We've used an alignment flowchart and identified what adds value to our product. We discovered that we want to target two different users with our product design.

Based on our analysis, our team decided it would add value to offer a "white glove" service that can assemble and adjust the bike stand for the user. This would help reduce or eliminate steps for the customer, making their use process easier.

What that means is that we have to design for two users: the "white glove" installer and the bike owner. A deployment flowchart helps us understand who is doing which step in our process.

Using a deployment process flowchart, we simplify the use process steps that don't add value (from our value-added analysis) and add a white glove service.

Let's see how this would look in a Deployment Process Flowchart.

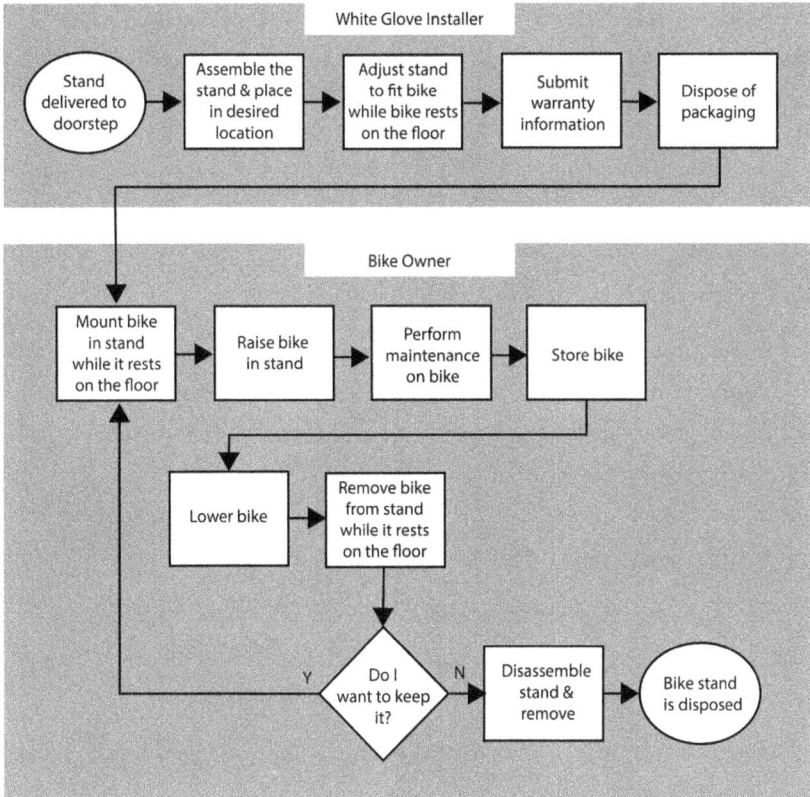

Figure 10.9. A Deployment Flowchart of the Running Example Scenario

Deployment flowcharts are really helpful when we have customers who may have different experiences that we want to target and different use scenarios. A flowchart can also be helpful to understand process steps that occur simultaneously. It uses swim lanes: one swim lane is dedicated to one type of user and another is dedicated to a different user. We still draw out our process flowchart, but now we segregate the process steps into who is doing what. That's how a deployment flowchart can help us better understand the use space and our users.

(For the ADEPT Team Framework activity steps for a Deployment Flowchart, refer to Appendix E.)

Iterate with the analysis to learn more, but don't overdo it.

In the scenarios in this chapter, you've seen that we can do multiple analyses on one use process. We started with an Alignment Flowchart analysis, then continued to explore it in the Critical to Quality analysis. We reevaluated our Alignment Flowchart analysis as a value-added analysis, then further analyzed it with a Deployment Flowchart.

It all comes down to what you want to learn about the use process. It is not necessary to do all these analyses. But if there is more than one question your team has about the use process, then go ahead and do the analysis you need to do.

The main idea is to build on what you've learned for concept development. Don't just do an analysis for analysis' sake. Have targeted questions for which you want answers or want to explore more deeply. Just like you wouldn't test a product without a plan, you wouldn't run a series of flowchart analyses without knowing why.

Below is a simple summary of the purpose of these analyses. I've tried to distill their power by using only one verb:

- Alignment Flowchart: clarify

- Comparison Analysis: change

- Critical to Quality Analysis: evaluate

- Value-Added Analysis: simplify

- Deployment Flowchart: synchronize

Choose how you want to take next steps with the use process. Then use the ADEPT Team Framework and these analyses to help you knowledge share with your team.

Make choices about design inputs from the use process analysis.

Using the information gathered with our cross-functional team, we can explore the use process and start drafting design inputs. We can continue to use it throughout the development process. Remember that we use co-working as a learning opportunity with our cross-functional team before we embark on engineering design.

1. **Update the user descriptions for this project along with their corresponding needs.** Your team may have uncovered new information during the exercise, so be sure you capture it in your user descriptions, especially so if you've used a Deployment Flowchart. This co-work session may also generate more questions for our customers.

2. **Share your insights with your team in a discussion.** The meeting is where you discuss your recommendations about the steps that are important to control by the design. In the meeting, you show how their co-work in evaluating the use process fits into design decisions. Keep the feedback loop going, along with an open mind, viewing this as another opportunity to learn what your teammates know.

3. **From the use steps that were newly identified and/or prioritized, develop design concept options.** What about the design concept could you create or change so these steps happen in the way the team has prioritized? Keep in mind the type of analysis you used to figure out how to treat the use step.

4. **From the analyses, develop concept offerings.** Offerings may exist outside of the design and include other functions of the business. What offerings of the product or service could enhance the steps that are a priority?

5. **Develop design inputs**. Maintain the information to iterate design throughout the development process. Identify steps that are critical to quality and value or design inputs that are associated with them.

Keep in touch with your team.

Maintain communication with your team about your design decisions as often as you can. Being able to link your design to their input helps ensure the best possible design for the user. And it eases the buy-in needed once the design is completed. It also continues to help you make design decisions about aspects of design that affect the use process.

Key Takeaways:

1. Flowcharts help teams **align** on the scope of the use process, clarify the users, understand differences and gaps, identify interface requirements, and prioritize what needs to happen for success.

2. There are five main ways to analyze a flowchart for design inputs: **alignment, comparison, critical to quality, value-added, and deployment.**

3. The flowchart and its analysis can be **iterated** as the team learns more about the product and its use, moving from high-level concepts to a more detailed analysis.

Reflection Questions:

1. Does my team know the steps that add value for the user and those that are critical to quality? Am I designing for those steps?

   _____

   _____

2. How can I use a flowchart to help my team better understand interfaces and inputs that affect the use process?

   _____

   _____

3. Does my team need to analyze a flowchart to understand how different users or groups interact with the product and each other?

   _____

   _____

4. When is the right time for me to make changes to a flowchart and move from a high-level view to more detail?

_____

_____

Part Two Review

Key Concepts

Here are some key concepts covered in part 2:

1. Use the Concept Space to align your team on the scope of concept development, better understand customers, and target customer experiences. Prioritize what you and your team want to learn in the next steps.

2. Break down benefits and symptoms to include impact and prioritize them based on impact to the customer. Use these targeted models/ templates to gather information about potential features, offerings, and risks to avoid.

3. Consider what you want to learn about the use process for design inputs. Choose critical to quality, value-added, or deployment analyses to better understand this concept idea.

Practice Activities

Here are some action-planning exercises to help you apply the concepts from part 2.

1. **Benefit Statement Practice**: List three intended outputs for a product design you are working on, then rephrase each as a benefit statement that includes the feature and its impact. Next, consider where each statement fits into the Kano Model and how the impact relates to an expected level of implementation.

2. **Symptom-Impact**: Find out if your group performs System failure mode effects analyses (FMEAs). If so, determine the most relevant severity rating scale. Can you use it to prioritize the impact of

symptoms you're trying to avoid? If not, what rating scale could you use?

3. **Use Process Mapping**: Think of a product you currently work on and map out its use process. Start with a clear beginning and end and fill in the steps and decision points. If it's helpful, you can also start by looking at the use instructions for a previous product or a competitor's product. This activity helps you identify areas where you need more information on the use process. It also helps you prioritize the steps for product design based on quality and value.

4. **Concept Space Understanding**: Do you have a good grasp of the concept space for your current project? Consider the following questions and write down your answers. You may use Figure P2.2 as an aid. Your assessment should take about 20 minutes. Success is understanding where you might have gaps in design inputs.

   4a. Conditions/Assumptions: Do you really understand the users, their assumptions, and the use environment so you can meet the customers where they're at?

   4b. Benefits:

   - Do you have a list of potential and target benefits, both stated and not stated by the customer?

   - Are they related to potential design features?

   - Are they ranked by the level of customer satisfaction they may bring? If your product delivers more of it, will customers like the product more?

   - What about the concept design relates to these high-ranked features?

   4c. Symptoms:

   - Do you have a list of potential symptoms, or problems the users may experience when something goes wrong?

   - Can you rank them based on the significance of their negative effects?

- What design inputs may lessen/prevent these problems?

4d. Use Process:

- Do you understand the basic steps users may take with your concept?

- Is your team in alignment with the use process?

- Do you know the steps that add value to the user or the steps that are critical to an experience considered high quality? Are you designing for those steps?

- Do you have multiple users doing different tasks, and which tasks are done by whom?

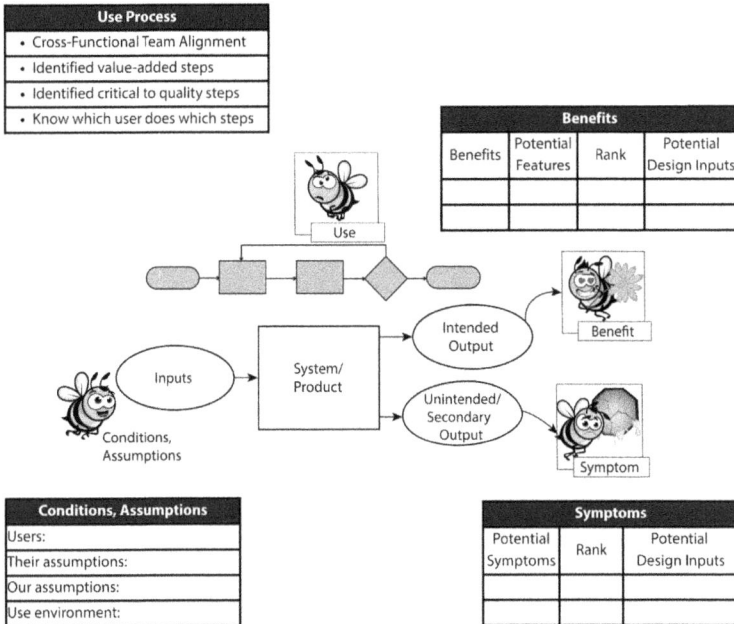

Figure P2.2. Concept Space Understanding Activity

Part Three

TRANSFORM SHARED KNOWLEDGE INTO WINNING PRODUCT DESIGNS

IN PART 3, we'll focus on information for design but also build on what we've learned in part 2 and move farther along the product development process.

I'll show you ways to:

- Improve the information from the Benefits-Impact Template with other analyses. **Tree diagrams** can help us develop vague drivers to design inputs. **Matrix diagrams** can refine and link customer benefits to design requirements.

- Study information from the Symptoms-Impact Template through a **Risk Analysis.** Also, import it into a **System FMEA**, and develop design inputs that eliminate, reduce, control, or otherwise mitigate potential risks.

- Take the Use Process Model a step further with a **Task Analysis** that's based on the perception-cognition-action (PCA) framework.

Now, we extend concept development into the design space, getting more into technical design inputs. We still haven't started engineering solutions, and that's often because we don't have a straightforward process for integrating our work thus far into later phases of development.

We gather information and work together toward a common understanding that helps us make decisions, but then the information sits dormant for the rest of the project. It's not carried through into development later because we think we've already incorporated what we learned into the immediate design decisions, but that's not the case.

One thing we don't want to do is settle into analysis paralysis where we hash out the same details repeatedly and do not actually move forward with product design.

Instead, we want to constantly iterate what we know and what the product will be. The product design at the end of a development project is rarely exactly as we envisioned at the beginning. Trade-offs occur. Or, during development, we learn something new about what we're designing and decide to change our minds. Risk analysis may point to a redesign that has better risk controls than our original concept.

We can refer to what we did in the beginning to ensure our customer information continues to be integrated into our design decisions, but a better way is to continually improve on those early concept ideas. We're going to pull information into our design decisions as we develop the product.

Concept development is just the start of the conversation.

It's not like we just forget about what we did during concept development. We continue the conversation with our cross-functional team through the rest of product development.

We haven't wasted or completely signed off the work our concept team did. We continue to use it to help us make decisions about the functionality and performance of the product itself.

Information from our concept development work can help us answer these questions:

- What are the priorities of these distinct features and capabilities?
- How many samples do we need to test?
- Where do we need to work with our suppliers to better control quality?

- What customer service capabilities do we need to change to best support our customers with this new product overall?

Use the results of the Concept Space Model analysis as a starting point for other analyses.

We can iterate what we've done for concept development throughout product development to answer more questions as we're developing the product. Consider:

- By mapping customer benefits to the requirements that serve them, we can help prioritize design inputs.

- When customers experience symptoms of a potentially bad event, we can analyze and use those symptoms in an FMEA (failure mode and effects analysis) for risk-based decisions and controls.

- We can use tools like the process flowchart in a task analysis to find use failures, so we can mistake-proof the design.

Our team is also going to use this information to support the project in other ways. We work with our team in the concept space to gather their design inputs, but it isn't just one sided. Your team has its own goals and gathers the same inputs for their own purposes, even as you all work toward the same goal.

Results from the Concept Space Model help us focus on what matters most.

We adjust our efforts on this project based on what's needed for success. If something is critical, we can do a deeper analysis with that feature.

We don't have to go as far with every feature. We can adjust our analysis based on what's important and what we need to learn. Customer experiences are prioritized as we gather information during concept development. Knowing these priorities helps us focus on what's important and gives us the information to make any necessary trade-offs with our assumptions.

The Concept Space Model technique applies to all types of designs.

If you have suppliers doing something completely different from you, the tools and methods in the Concept Space Model can still help you establish common ground. Also, the Concept Space Model is transferable between industries. Whether you are making office products, medical devices, or developing services, you can use the same techniques for concept development with your cross-functional team.

The ADEPT team meeting framework can be adapted to other team activities.

Continue using visual models/templates and the ADEPT teamwork model to do these other activities with your team. You do not need to do them all. Decide what it is you must learn more about to make decisions about design inputs.

Continue to refine design inputs with analyses, as needed.

The table shows how information we gather at concept development progresses into more specific design inputs. The methods included in this book are shown in bold.

Table P3.1. Possible Progression of Concept Development Information

First Analysis at Concept	Possible Next Analysis after concept - design form	Possible Further Analysis - design form refinement
Concept Space	Schematic with blocks and fundamental interactions	• Geometric layout with physical implementation, layout, and human interfaces • Incidental Interaction Diagrams, prompted from a Geometric Layout
Benefits Analysis with Customer Satisfaction	**Tree diagram** to develop vague drivers to design drivers	**Matrix diagram:** combination L and roof-shaped matrix.
Symptoms Analysis with Severity	• **Risk Analysis** • **System FMEA**	• Design FMEA • Process FMEA • Fault Tree Analysis
Process Flowchart	**Task Analysis** using a PCA (perception-cognition-action) framework	• Poka-Yoke, mistake proofing • Usability FMEA

I'll show you ways to pull information we gathered from part 2 with our teams into other analyses for design inputs. However, this book does not cover how to draw schematics, layouts, and interaction diagrams from the concept space. I recommend them for continued team input and refer you to *Product Design and Development* (6th edition) by Karl T. Ulrich and Steven D. Eppinger.

Let's continue working with the Concept Space Models.

Chapter Eleven
TRANSLATE CUSTOMER VALUE INTO CONCRETE DESIGN

"The value of an idea lies in the using of it"

*- Thomas Edison, whose relentless pursuit of
practical inventions reshaped society*

———⬦⬦⬦———

IN THIS CHAPTER, we highlight how the drivers to benefits are the beginnings of design inputs. We'll assess drivers to benefits for design inputs. Then, we'll construct tree diagrams of benefits to design inputs. Finally, we'll construct a matrix to compare features with design inputs.

In part 2, we used a Benefit-Impact Template with which we explored features and their impact. We added drivers and applied a customer satisfaction rating to the impact so we could prioritize our design features.

In this chapter, we're going to take what we've learned and continue to develop that knowledge into technical design inputs. We can explore the drivers that aren't quite to the level we want or can design against.

A tree diagram is a simple tool that can help us get to that point with our team. Next, we can use a combination of two matrices to compare the characteristics we've been developing against themselves and the benefits. Are we really developing the right design inputs? Which ones are the most important for quality, reliability, and cost?

Use a tree diagram to help your team develop drivers into design inputs.

Let's look at those drivers to our benefits and see how we can get design inputs from them.

We've used the Benefit-Impact Template to create opportunities and increase positive experiences. We look at what can drive the feature to be available and what can drive the impact given that the feature is available.

With all drivers, we're looking for design inputs. In our previous scenario, we came up with different drivers for features compared to the impact.

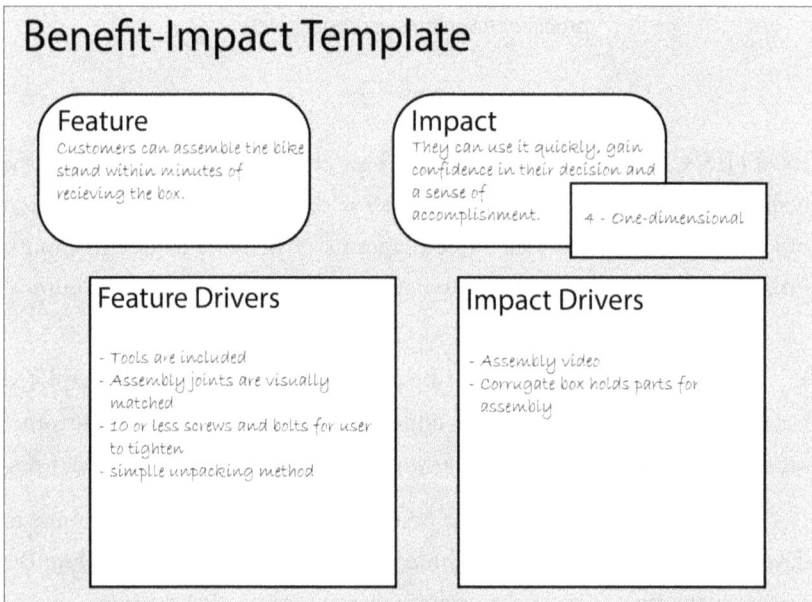

Benefit-Impact Template

Feature
Customers can assemble the bike stand within minutes of recieving the box.

Impact
They can use it quickly, gain confidence in their decision and a sense of accomplishment.

4 - One-dimensional

Feature Drivers
- Tools are included
- Assembly joints are visually matched
- 10 or less screws and bolts for user to tighten
- simplle unpacking method

Impact Drivers
- Assembly video
- Corrugate box holds parts for assembly

Figure 8.9.

234

Drivers of the features are included tools, visually matched assembly joints, a limited number (10) of screws and bolts, and a simple unpacking method. Some of these inputs are well on their way to being developed, and some of them are not quite at the level of a technical design input.

An activity we can use to help us get to design inputs is a tree diagram. Tree diagrams are a useful model. Many tree diagrams are likely familiar to you. There is the quality why-why diagram. Fault tree analysis is a type of tree diagram. Even the organizational charts your company uses are a type of tree diagram. A tree diagram helps us map out vague drivers for features into design inputs.

Example Scenario

In our bike stand example, we started with the vague driver "simple unpacking method." What could that mean? Two approaches address this characteristic: "parts are unpacked from the box(es) in order of assembly," and "the user can access the assembly instructions before unpacking parts." We continue to break down the characteristic until we get to an answer.

Our design inputs are meant to achieve a specific impact on the customer. For our example, there are design inputs that contribute to the feature "customers can assemble the bike stand within minutes of receiving the box." By having parts layered in the box in order of assembly and making assembly instructions available before the customer unpacks parts, we contribute to features that help our customer use the product quickly, gain confidence in their decision to buy the product, and have a sense of accomplishment upon its assembly.

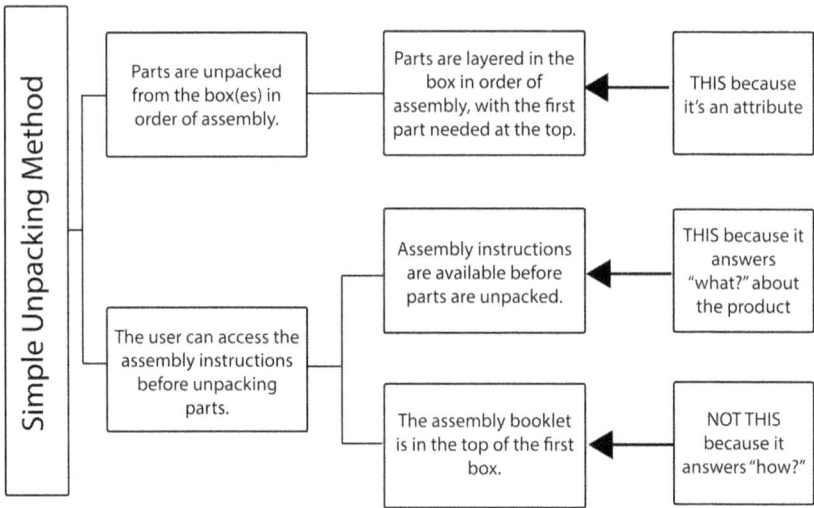

Figure 11.1. Tree Diagram, developing a feature driver into technical design inputs

When mapping out a tree diagram from driver to design input, we don't want to get too specific, too fast. We also don't have to provide the same level of detail for each branch. We just want to go far enough so we can define a characteristic that's necessary and sufficient to be a proper design input. You may define characteristics that are structural, behavioral, or logical. They may be concrete or abstract objects.

Industries, companies, and products may have different goals for design inputs depending on their design control process. Your final characteristic might be a requirement or a metric. It may be a property-based requirement—for instance, "When the relative humidity is 60%, the reliability of the switch to successfully close the circuit shall be 85% with 90% confidence after a dormant time period of 300 days."

Use your best judgment in defining a design input for your product. Rely on what you learned in concept development to link those design inputs to customer experiences. And make trade-off decisions appropriately based on the priorities you learned about in concept development.

Combine benefit features with design inputs using matrices.

We may have used a tree diagram to develop design inputs. If our Benefit-Impact Template with drivers is straightforward, we may not need a tree diagram. In either case, we now have a list of benefits, characteristics, and their relationship to customer satisfaction. We can better understand the relationship between those three elements by using a matrix diagram.

There are two different matrices that work well in this situation. One is an L-shaped matrix, which compares two groups of unique items. The other is a roof-shaped matrix, which compares one group of items against itself. Combined, the L and roof-shaped matrices can really give us some important insights into our customers and what it is we're designing.

The matrix resembles the House of Quality used in Quality Function Deployment (QFD), but we won't develop it as extensively as QFD does due to conflicts with other design processes. Additionally, QFD needs to be intentionally implemented throughout the organization. For our purposes, we will only use it to develop our design inputs.

In this chapter, we set up these matrices to look at the relationship between features, design inputs, and customer satisfaction.

People use the combination of an L-shaped matrix and a roof-shaped diagram to:

- Ensure our features aren't being duplicated.

- Ensure we have technical design inputs that enable the features that matter.

- Prioritize our list of design inputs by asking: Which inputs are associated with what level of customer satisfaction? What is the strength of the association between inputs?

- Understand how to link design inputs for tradeoffs and identify the ones that support more than one feature.

- Refine design inputs according to similar groups, eliminating those that are unnecessary, and restating inputs to clarify them or be more succinct.

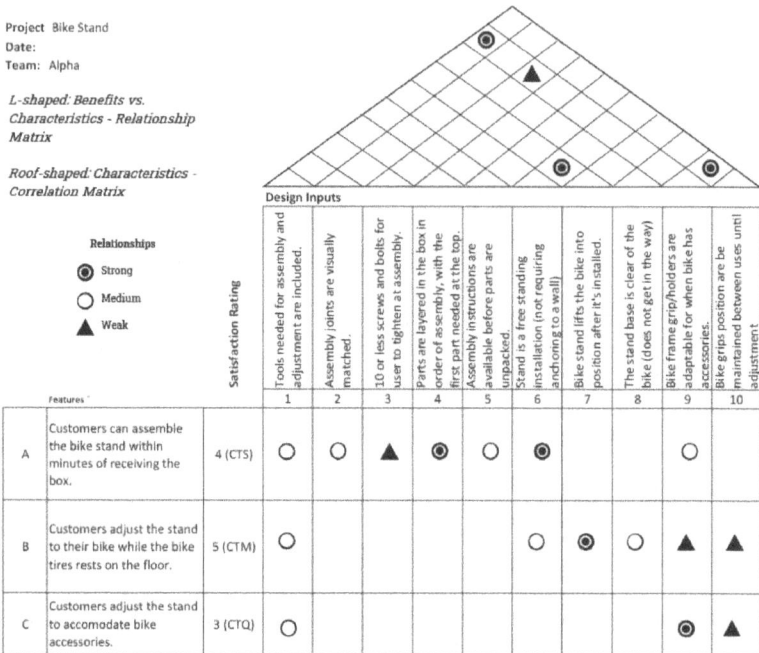

Figure 11.2. The layout of a combination L-Shaped Matrix and Roof-Shaped Matrix to connect desired features to design inputs

Here are the steps to construct and use this matrix diagram:

1. In the L-shaped matrix, list the features by row, ordered by customer satisfaction priority. The priority could be a customer satisfaction rating that's associated with the Benefit-Impact Template.

2. In the columns, list our drivers and/or design inputs.

3. Evaluate the strength of the relationship between features and design inputs, categorizing it as strong, medium, or weak in the intersection of rows and columns. This area of the matrix is called the "relationship matrix."

4. In the roof-shaped matrix, using the same relationship legend, identify the design inputs that correlate or compete. This area of the matrix is called the "correlation matrix."

5. Examine the results to make choices.

Draw a matrix diagram with shared tools.

Creating a matrix diagram is something that can be helpful to do with our team. Let's create a simpler scenario for ways we would evaluate a matrix diagram.

Here's an example matrix I put together using an online whiteboard. We have our L-shaped matrix in rows and columns. I used yellow Post-it notes to represent features, from our Benefit-Impact Template. Those are placed along the left to make rows. I used pink Post-it notes to represent the drivers from our Benefit-Impact Template, or the design inputs. They make up the columns in our matrix. I used a pen to draw out a roof-shaped matrix above the drivers.

We also have symbols for the relationships we're comparing. We have strong, medium, and weak relationship measures. Our team discussed these relationships and marked them on our matrix. Now, let's look at what we can learn from this matrix.

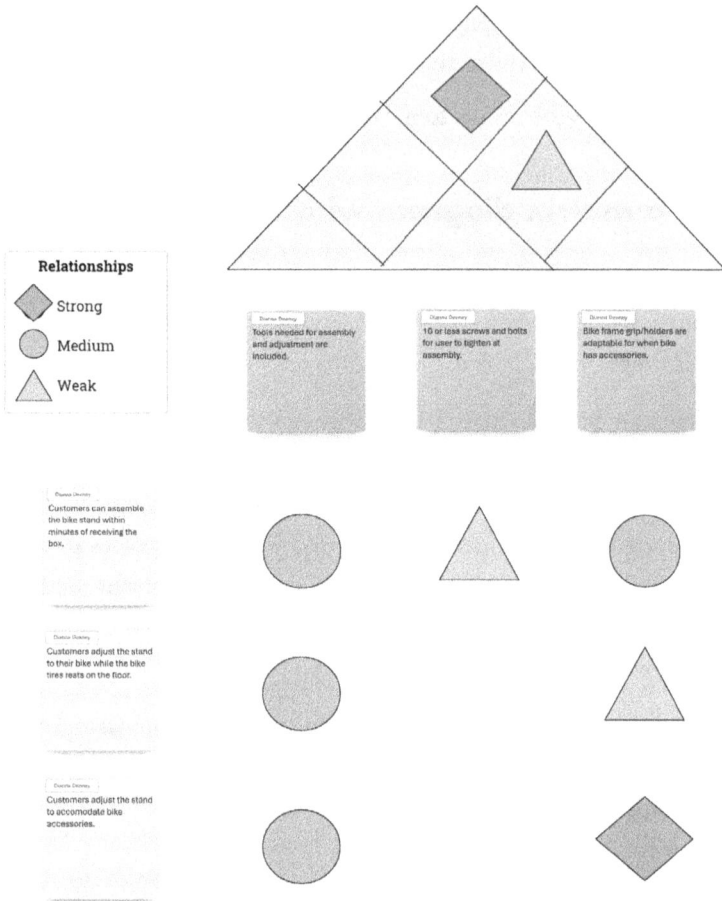

Figure 11.3. A layout of a matrix on a shared digital whiteboard with digital sticky notes and shapes

Examine rows.

Just looking at the L-shaped matrix and ignoring the roof-shaped one for now, we notice that no two rows are the same, so we don't think features repeat. No rows are empty, which is also a good thing. It means all our features and benefits are being represented by a design input.

There are two features with no strong relationships to an input, and this could be a problem. Each feature should have one strong design input. This might show that we need to revise the input or maybe add another one.

Examine columns.

Let's look at the columns of our L-shaped matrix. There are no empty columns, so there are no design inputs that aren't fitting a feature. The first characteristic shows a medium relationship with all our features. The third characteristic, however, exhibits different types and strengths of relationships with each feature: strong, medium, or weak for each of our three features.

Examine the roof.

Consider the roof-shaped matrix and how these inputs relate to one another. We see the first and third design input have a strong relationship, and the third also has a weak relationship to the second.

Examine the matrix overall.

Reviewing the matrix overall, we can see the third design input relates to all the features and the other inputs. That means this input is going to be a priority for our design.

We need to ensure customer satisfaction, particularly with strong relationships. This design input might be an issue for reliability, safety, or cost. We'll want to investigate this design input more thoroughly to understand what's happening and how it could affect the development and performance of our product.

Is the first design input necessary? It has a medium relationship to all the features and is strongly related to the third. Does the third design input supersede it? We may have an unnecessary input, which is something we should investigate further. If it's unnecessary, we can remove it and simplify our project.

Make design input decisions with the information in the matrices.

As you can see, these matrices together help us evaluate our features and design inputs, including the quality and cost of what it is we want to implement. We can make sure our inputs are covering our features.

We can prioritize inputs by considering both customer satisfaction ratings and the strength of associations. We can understand how inputs link so we can see which ones are competing for trade-offs. If there are certain inputs that support more than one feature, there may be quality and cost concerns.

These matrices can also help us refine the design inputs. Can we group some if they're similar? Can we eliminate others because they're unnecessary? And are there other ones that need to be clarified? These are all the things we can talk about with our team using the L-shaped matrix and the roof-shaped matrix together.

(For a checklist to aid in analyzing these matrixes, refer to Appendix F.)

Avoid common pitfalls when using matrices.

Creating a matrix diagram is something that can be helpful to do with our ADEPT team practices. You can imagine matrices getting big and unwieldy. For matrices at concept development, 25 items are too big. We really want to focus on the high-value benefits or the benefits our team finds confusing and we need to explore more.

Another problem teams have with using these kinds of matrices is that they linger on the details past the point of learning for action. They have "analysis paralysis." To avoid this, we really want to focus on teamwork and learning for decisions, not perfection. We want to take something, understand the relationship, understand where we may need to do a little more research, and then move on.

These matrices, for example, can highlight a problem of conflicting customer bases. Maybe you're developing one product, but you really have two types of customers who would use that one product. That could lead to confusing results. To address this, you could use a matrix for each of your customer bases or prioritize one type of customer over another. Or you can average out the satisfaction level for a benefit.

However you choose to handle that situation, be consistent with your project. When creating these matrices with a team, use a shared space, like Post-it notes with a whiteboard. Don't use a spreadsheet. It looks nice, but it's terrible for team dynamics during concept development as we explained in part 1.

Just like other ADEPT methods, we want to align our team with our prioritization method. We can use the same relationship legend as in the example to identify strong, medium, and weak relationships. Sometimes teams decide to use a different legend, one that shows a positive correlation or a negative correlation. You can decide with your team what would work best for you. If you're starting out with something like this, I recommend sticking with one legend for both the L-shaped matrix and the roof-shaped matrix.

We could add more rows and columns of information to these matrices, such as the level of difficulty to implement a feature. Some teams add the measurement unit or the target spec. That type of information could be useful to your team to make decisions.

Key Takeaways:

1. Drivers to benefits are the starting points for design inputs.

2. Tree diagrams are useful tools for mapping vague drivers for features into design inputs, breaking down characteristics until you reach attributes that define "what" about the product. These design inputs should contribute to a specific impact on the customer.

3. Matrices (L-shaped and roof-shaped) shed light on the relationship between benefits, characteristics, and customer satisfaction. The L-shaped matrix compares features and design inputs, while the roof-shaped matrix compares design inputs against each other.

4. The combination of L-shaped and roof-shaped matrices can help ensure features aren't duplicated, identify technical design inputs that enable important features, prioritize design inputs, link design inputs for tradeoffs, and refine design inputs.

5. Teams should avoid "analysis paralysis" and focus on teamwork and learning for decisions when using matrices. Matrices should focus on high-value benefits or benefits that need more exploration.

Reflection Questions:

1. Looking at my current project, how have I translated the drivers of customer benefits into concrete design inputs?

   _____

   _____

2. Considering my design inputs, how well do they link to my customer's experiences and priorities as I understood them in concept development?

   _____

   _____

3. When mapping out design inputs, did I use tree diagrams to break down vague drivers into more specific, actionable characteristics? Did I go far enough to define a necessary and sufficient design input?

   _____

   _____

4. When using matrices, did I avoid "analysis paralysis" and focus on teamwork and learning for action? Or did I get stuck in the details? What did I learn from the matrix analysis that I might not have seen otherwise?

   _____

   _____

Chapter Twelve

ENGINEER RESILIENCE THROUGH SYMPTOM-DRIVEN DESIGN

"An ounce of prevention is worth a pound of cure."

- Benjamin Franklin, the quintessential American innovator and nation builder

———◦⟨⟨⟨⟨⟩⟩⟩⟩◦———

IN THIS CHAPTER, we take the next steps with the information from our Symptom-Impact Templates. We're still driving toward design inputs. We assess symptom risks to gather information and prioritize design actions. Then, we put information from our Symptom-Impact Template into an FMEA for risk analysis.

In concept development, we do not have specific problems or causes to examine. But our team can identify potential symptoms our customers may experience.

In part 2, we explored possible symptoms by breaking them down into outcome and impact and listed drivers and assigned each a severity rating. We showed that closely looking at the symptoms can reveal many additional factors to consider. Breaking them down into their parts helps us and our team *really* get into some interesting things that can help form our design inputs for *great* designs.

Having reached that point, we can continue to iterate what we've learned through the development process.

Assign probabilities to prioritize design inputs further with a risk analysis.

Dividing a symptom in two helps us to think better about the likelihood of something happening. Thinking solely about the outcome helps us grasp its probability more easily. And we can more easily think about the probability of the impact once the outcome has occurred.

This model enhances our thinking even more when we consider the drivers. A better understanding of the drivers leads to more ideas about how to address symptoms. This helps us guide our team to understand, prioritize, and rate risk events.

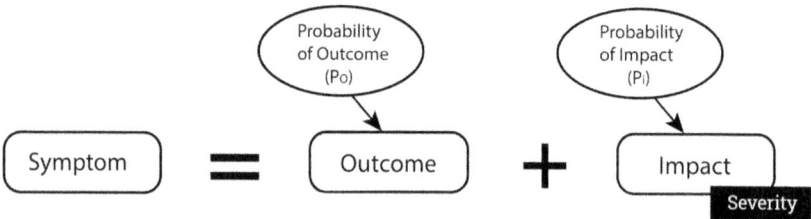

Figure 12.1. Adding Probability to the Symptom-Impact Model

We want to expand the Symptom-Impact Model to evaluate the risks of a product concept. In this situation, we are asking our cross-functional team to assign a probability based on their experience and knowledge. There is no expectation that we will have test data available. The questions we ask are:

- **What is the Probability of the Outcome?** What is the likelihood this outcome will occur? Consider the drivers of the outcome to help you assign a probability.

- **What is the Probability of the Impact?** What is the likelihood this impact will occur, given that the outcome has happened? Consider the drivers of the impact to help you assign a probability. This is a conditional probability, meaning that it happens only because of another event.

We can combine these probabilities to understand the likelihood of the symptom. To do this, we use the rules of conditional probability: The probability of the symptom equals the probability of the outcome multiplied by the probability of the impact, or:

$$Ps = Po * Pi$$

(For more details, refer to Appendix G.)

Assess symptom risks to gather information and prioritize design actions.

Let's analyze the symptom risk so we can prioritize activities.

We can assign a risk priority number to a design function or feature based on symptoms. This is not the classical RPN from FMEA (risk priority number of failure mode and effects analysis). It's just a priority number we assign to a risk we evaluate.

We can create a simple table that captures our symptom with its two different parts: the outcome and the impact. We can also capture the two different probabilities. One is the probability of the outcome. The other is the probability of the impact once the outcome has occurred. We also can capture the severity of the impact.

For our example, I created a simple severity rating scale that's numerical, but it's really based on quantitative information. Note that the severity we're going to list on our table is based on the impact. There are two calculations we can add to our table. We can add the likelihood of the symptom, which is a product of the two probabilities. Risk priority equals severity times likelihood.

Table 12.1. Compare and prioritize risks in a project with likelihood, severity of impact, and a risk priority.

ID	Symptom		Probability of outcome (P$_O$)	Probability of impact (P$_I$)	Likelihood of Symptom (L$_i$ = P$_O$ x P$_I$)	Severity of Impact (S)	Risk Priority (S x L$_i$)
	Outcome	Impact					
B1	Customer cannot assemble the stand.	It delays usage and takes time and effort to resolve.	0.2	1.0	0.20	4	0.80
B2	Stand does not fit bicycles with XL frames.	Customer uses the bike stand, but it loses some safety features.	0.8	0.6	0.48	6	2.88
B3	Customer must remove attachments to use the stand.	Customer needs to permanently modify bike attachments to fit stand.	0.6	0.8	0.48	3	1.44
B4	Bike stand's soft surface is damaged.	Customer uses the stand as-is but notices the defect as poor quality.	0.4	0.9	0.36	2	0.72

In our example, we saw the customer cannot assemble the stand, which delays their ability to use and takes up their time to resolve. Our risk priority measure is .80. This doesn't really mean a lot on its own, so let's use three other scenarios to compare them.

I developed these scenarios using bike stand review data. Our customers face these scenarios when using competing bike stands. In our scenario, the lower the risk priority number, the less risk our symptom carries.

One scenario includes a symptom with an outcome that "the stand doesn't fit bicycles with extra-large frames." The impact is that "the customer uses the bike stand but it loses some safety features." Our team evaluated this outcome and impact combination and assigned a severity of the impact. Our risk priority for this symptom is a 2.88. We have two other scenarios listed in our table and have calculated risk priority numbers to be 1.44 and .72.

Remember, we're not evaluating any particular problems or causes. We're looking only at the high-level symptoms—things that could happen or that our customers could experience if there is a problem. Considering that this often involves a chain of events, we're really focusing on the first things we notice when something's wrong so we can better understand the riskiest design features or understand what needs further follow-up.

We could use the risk priority alone and prioritize our activities and our design inputs based on that number. But we might miss out on some other key factors, such as likelihood and severity of impact. For a more complete comparison, we can map risks on a scatter plot: The X-axis shows severity. The Y-axis shows likelihood.

To continue our example, I also created a scatter plot of our different symptoms. In this scenario, we clearly see that B2 has the highest risk priority number. The likelihood of occurrence looks to be about equal to other risks. We prioritize B2 development within our concept.

We also see B3 is something we should focus on. It has a high likelihood of occurrence, or at least as high as the earlier symptom we noticed. When we're talking and working with our cross-functional team, we may decide these two symptoms are a top priority. If we don't get these things right, then we won't have happy customers or a quality bike stand.

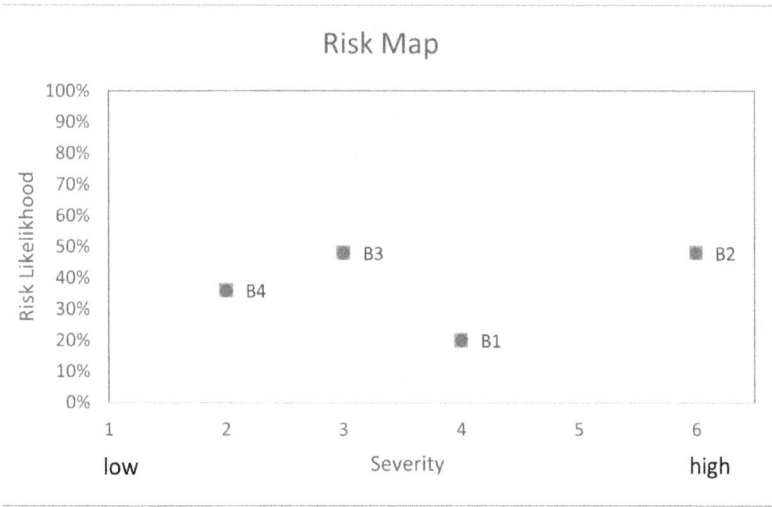

Figure 12.2. Plotting the symptoms onto a scatter plot for comparison

Given this information, we also have next steps. Our user needs should include that the bike stand is suitable for extra-large bicycles. Or we need to clarify that our bike stand is suitable for bikes up to a particular size. Do we need to limit our customer base or design it for a bigger customer base?

We also need to consider the type of bike the customer is storing. Is it a cruise bike with fenders? They may not fit in these stands unless the fenders are taken off.

What do we know about the latest bike accessories, like radars, lights, bike racks, and bags? These are the things that require top priority.

If we don't get these things right, then we won't have happy customers or a quality bike stand. We want to talk with marketing and field operations to understand more about our customers. How many of them use extra-large bikes, smaller bikes, cruisers, or electronic bikes? What types of accessories are they using daily?

We can do our own technical investigations into how bikes fit on the stand, considering factors such as wheel size, thickness, and frame size. How do different bike sizes compare? And when we get a better understanding of accessories, where are they placed on the bike? Which bike parts should the stand avoid?

By looking for potential symptoms, we can uncover what problems our customers may face. As we've seen with this example, we develop more specific questions that we can ask our cross-functional team or the customers themselves. We have a starting point for technical information needed for design. And we also have information about what's important to this design and which features may be more important than others.

We want to keep the risk priority exercise simple. Remember, we use it to inform our decisions about what's most important for the design and for our customers. We're going to use this information to help us develop design inputs. We can use the Symptom-Impact Template to understand outcomes and their impacts, which helps us capture and prioritize our early activities.

Push the Symptom-Impact Template results into an FMEA for more design inputs.

FMEA (failure mode and effects analysis) is possibly the most popular analysis tool for risk management. Different standards and regulations about risk management rely on FMEA as the primary tool, and there are different standards for various industries.

Consider:

- If you work in automotive or as a supplier of automobile components, you'll likely follow the "FMEA handbook" published by the groups Automotive Industry Action Group (AIAG) and the German Association of the Automotive Industry (VDA).

- If you work in the medical device industry, you'll follow the ISO 14971 standard.

- If you work in civil aeronautics, you may need to comply with ARP-4761 issued by SAE International.

The list goes on!

My point is that FMEA is a universally recognized, useful analysis that teams are expected to use during design development. A completed FMEA table also shows some evidence of compliance against these standards and regulations.

A common mistake occurs when an FMEA is done as a checkbox item. With a bunch of regulators telling you that you need this document, it's easy to think, "Let's just put it together and get it on the books." But this is a waste of a fantastic analysis tool!

Our team is missing out if we don't approach FMEA as an input tool. Waiting until we have the design finished first is the wrong approach because it's being implemented far too late in the design cycle. This only makes it more likely that we'll be worrying about problems after our product is already designed.

Now, let's get into the real reason why we do FMEAs and how it fits into the design process. FMEA is traditionally considered a reliability tool, but some people also look at it as a quality tool.

Generally, how it works is that a team of people get together, talk about the potential ways things could go wrong, give values to those ideas, then assign actions or activities to eliminate or control risks. FMEA is a process a team can use to organize information about risks. It can also help team members decide if the action they took to reduce a serious risk was enough to reduce or eliminate that risk or if more work is needed.

Some uses of FMEA for design development:

- We continuously improve our design concept early on, making risk-based decisions as we hash out details about the design.

- During the design process, we can use it to help us figure out tests that need to be performed and define confidence levels for our tests as well as other controls, like detection controls.

- Late in the product development process, we can use it to make a final benefit risk decision.

- After our product is released in the market, we use FMEA to compare what we're seeing in the field against how we designed the product, and we continue to evaluate any performance risks.

Think of FMEA as an information tool. If our product design and monitoring system were a conveyor belt, then FMEA is a way to continually measure risk and adjust our decisions.

Our symptom statement, from the concept space, is the start to a System FMEA. FMEAs are driven first by whatever function our product is supposed to perform. Then we evaluate a failure of that function. We also evaluate the potential effect of the failure on our customers/product functionality/overall procedure, however we prioritized the severity of that effect.

In the Concept Space Model, we focused on what our customers could experience when something goes wrong. Here, we are evaluating a potential outcome with our concept product and the impact it has on our customer.

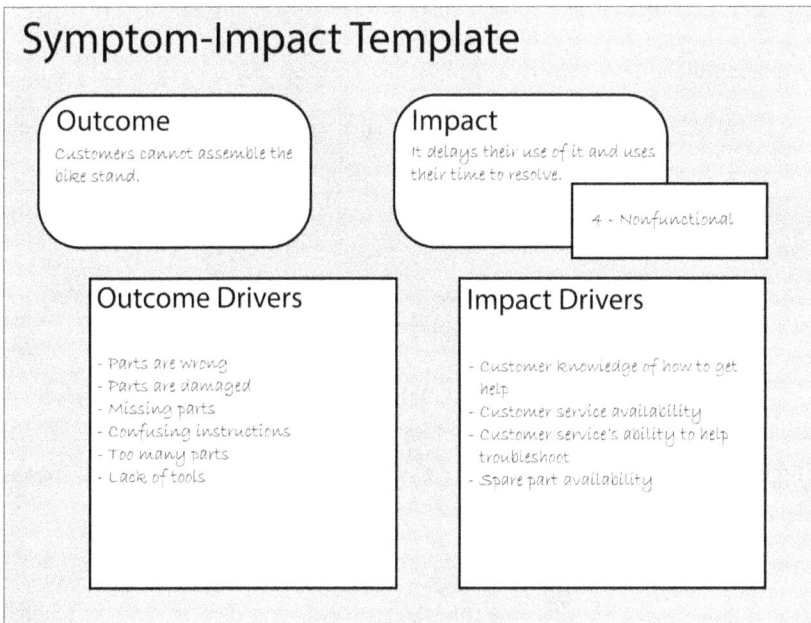

Symptom-Impact Template

Outcome
Customers cannot assemble the bike stand.

Impact
It delays their use of it and uses their time to resolve.

4 - Nonfunctional

Outcome Drivers
- Parts are wrong
- Parts are damaged
- Missing parts
- Confusing instructions
- Too many parts
- Lack of tools

Impact Drivers
- Customer knowledge of how to get help
- Customer service availability
- Customer service's ability to help troubleshoot
- Spare part availability

Figure 9.2

Our system function is the opposite of our symptom outcome in most cases. We may consider the outcome itself as a failure, the impact a potential effect to our customer, and the outcome drivers as causes. If we related our customer satisfaction rating scale to our FMEA scale, we could also bring over that data.

Table 12.2. Information from the Symptom-Impact
Template populates a System FMEA

System Function	Potential Failure Mode	Potential Effect(s) of Failure	Severity	Potential Cause(s)
Assemble the bike stand.	Customers cannot assemble the bike stand.	It delays usage and takes time and effort to resolve.	'Severity rating'	Parts are wrong Parts are damaged Missing Parts Confusing Instructions Too many parts Lack of tools

From an FMEA, we evaluate and prioritize risks for controls. Controls are both prevention (e.g., we designed-out the failure, made it less likely, or made it easy to detect) and detection (e.g., we can use these methods to catch the failure before the customer even sees it, or we use our design to detect the failure and stop the process). From these examples, you can see how design inputs are created from a System FMEA.

Creating and using an FMEA is beyond the scope of this book. I encourage you to find ways to incorporate the Symptom-Impact Model into FMEAs so you can continue to evaluate and prioritize for risk-based decisions and risk management.

There is little sense to starting an FMEA from scratch if you've used the Symptom-Impact Model. Pushing that information into FMEA only helps you gain more knowledge for better decisions. It also shows your team how their work is developing into design inputs.

Key Takeaways:

1. Breaking down symptoms into their outcomes and impacts helps to better understand the likelihood of a problem occurring. Probabilities can be assigned to both the outcome and impact and then combined to understand the likelihood of the symptom using conditional probability.

2. A risk priority number can be assigned to a design function based on symptoms, calculated by multiplying the likelihood of the symptom by the severity of the impact. This number is only one way to compare design functions with each other within the same project.

3. Mapping risks using a scatter plot can help visualize and prioritize the symptoms or design features that need further follow-up.

4. FMEA is a tool for risk management that can be used to continuously improve design concepts early in the design process. The information captured in the Symptom-Impact Template is a starting point for a System FMEA.

Reflection Questions:

1. Did I use the symptom information to generate more specific questions about my customers?

   _____

   _____

2. In what ways does the Symptom-Impact Template help me assign probabilities? How can I best involve my team in assigning likelihoods?

   _____

   _____

3. Do I currently use System FMEAs in my design process now? How can I connect the Symptom-Impact severity rating scale to my FMEA scale?

   _____

   _____

4. How can I ensure that risk management activities inform design decisions early in the product development process?

   _____

   _____

Chapter Thirteen

ELIMINATE USE ERRORS AND CREATE USER-FRIENDLY PRODUCTS

We must design our technologies for the way people actually behave, not the way we would like them to behave.

- Don Norman, pioneer of User-Centered Design and cognitive scientist whose work revolutionized our understanding of how people interact with products

———⟨⟨⟨⟨⟨⟩⟩———

IN THIS CHAPTER, we look at a definition of use error as it relates to product use and product design. This chapter also covers the three basic interactions users have with our product and how we analyze tasks and use errors for design inputs. Finally, we relate task analyses with other product development activities.

In part 2, we explored the use process for design inputs. We explored the general use process of how our customers get from A to B. We also explored concepts with our cross-functional team. Our vision of our product users is getting clearer along with the use environment and the use process. We're also understanding a little more about our product.

As we develop concepts, we come up with ideas that are going to involve a human interface. We want to evaluate that interface so we can ensure we reduce use errors.

A process flow also shows whether certain tasks might be more critical than others, especially where the user interacts with our product.

We'll explore more how to analyze those tasks from the user's perspective: what they perceive, what they know, and what they decide to do. In this way, we can mistake-proof our design.

Again, we're looking to design in controls and design out problems wherever possible. This results in a product that is the user-friendliest product we can make.

A use error deviates from what we expect.

By this point, we're used to thinking about potential failures and problems. A use error is really a problem or a failure we can design to address. It's a deviation based on what we expect to have happened or from what the user expected. It's not caused solely by a device failure but could result from a user action or inaction. If we can think about the use error as a deviation from what we expected the user to do, we're well on our way to better using the definition of use error for design inputs.

There are three types of use errors for design inputs.

Psychologists, human factors engineers, and usability engineers break down use errors into different levels and categories. For our design inputs, we're going to look at three high-level types of use failures.

They are:

- **Inadvertent action:** An inadvertent action is when our users know what to do but accidentally didn't do what they meant to do.

- **Thinking error:** A thinking error is an error of judgment. They did the wrong thing, believing it to be the right thing. You can also think of a thinking error as a mistake.

- **Deliberate deviation:** A deliberate deviation is when users deliberately do something the wrong way because it's the usual way. An example is speed limits on highways: Most people drive faster than the speed limit, though it's against the law. A deliberate deviation could be situational or because the task was difficult to impossible, so they skipped it or felt they had to do it a different way. Or they simply took a calculated risk.

Use errors happen because of the use process.

A key aspect of product design regarding use errors is recognizing that the error isn't because of the user. The use error happened because of the use process. The use process didn't adequately account for that kind of use error.

Note, too, the term "use error." It's not "user error" or "human error." It's a "use error," linked to the root cause of what is wrong with how our device works. We can't fix users, but we can design products to avoid errors.

Take a push lawnmower, for instance:

Figure 13.1. A Push Lawnmower Prevents Use Errors by Design
photo by senivpetro / Freepik

With a push lawnmower, you start the motor and push it along the ground to mow your grass. Well, sometimes people decide they want to trim their hedge with it. So, they reach down with both hands and pick it up and try to trim their hedge with the push lawnmower.

Naturally, serious injuries occur. That is a use error kind of problem. But if the product designers simply said, "Well, that's just the user's problem," and did nothing about it, that would be a design opportunity lost—and a future hazard.

The product designers added an extra switch—a lever that users must pull up and hold against the push handle—to fix that use error. If users let go of it, the motor stops.

The extra switch prevents the user from picking it up while it's running to use as a hedge trimmer. There's nothing to say the user couldn't tie the switch to the handle, but at least a built-in feature of the product is designed to prevent this use error.

Those are the things we want to look for in our own product designs. There are lots of benefits in doing so. It makes our products easier to use. Designing out use errors also decreases the support costs and complaints, and most important, it reduces safety risks. It also makes our product more desirable.

If we design for the users and design for use errors, our product might not only be safer but also more fun for our customers to use.

There are three task requirements.

We also have basic user interactions with our product. During a task, we can think about our user and our product or system as having an interface.

When our product gives an output, our users perceive that output with their senses. Then they process that information. Finally, they decide how they're going to act. That could be an input into our product design.

This is a standard framework, called the PCA framework.

Perception - Cognition - Action

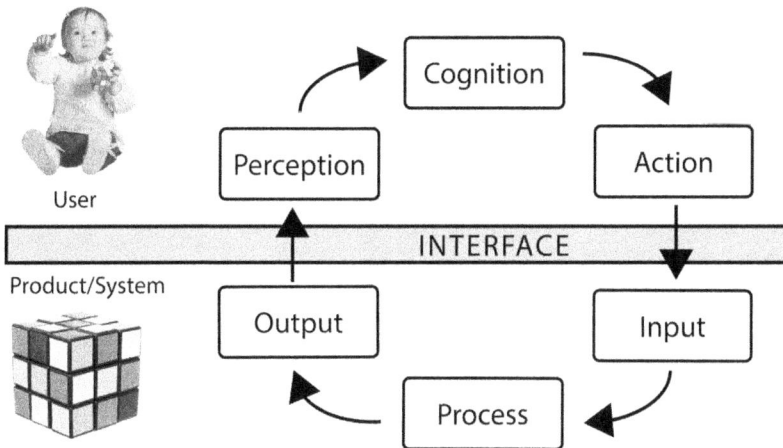

Figure 13.2. The interface between our product/system and the user
child photo by prostooleh / Freepik, toy photo by strefa / Freeimages

We can adopt the PCA framework as three requirements for every task or function we ask our user to take with our product. If the users can't meet a task requirement, a use error is likely to occur. We can think of PCA as the cause of the error. Users either don't perceive something, or they don't think about something in the right way, or the action they take wasn't correct.

> **Perception:** What did I perceive, or not perceive, with my senses that led me to make an error?

> **Cognition:** I apply what I know. What is my knowledge gap or misunderstanding of this task that led me to make an error?

> **Action:** What manual action did I take that led to an error?

We can use the PCA framework to figure out the root cause of a use error that has been detected by a test.

For example, later in development, we conduct formative studies during which we ask users to test our product and give feedback. In those studies, we may see some use errors.

We can use the PCA framework to get to the root cause of why that error occurred. Was it due to something they perceived? Was it a misunderstanding, or did they take the wrong action? We can also use the PCA framework to explore use errors for design inputs. We'll create design inputs to eliminate or control causes.

Develop design inputs from PCA.

We can develop some design inputs with the PCA model. What do our users need from our design to eliminate or control use errors?

Perception design inputs focus on information that users need to receive.

If our task needs the user to notice something, and they make a mistake, it's because they didn't see it. We can then think about design inputs as reducing or removing distractions or interruptions so they don't miss important steps or other information.

Mistake proofing is another option, which involves some kind of inspection, automatic or otherwise, to make sure the user or product does the task properly.

Alerts, alarms, or warnings can also show something is wrong. A great example of this are credit card readers. If a user leaves their credit card in the reader for too long, it starts to beep and holler for them to remove the card.

Cognition design inputs focus on how users understand tasks.

If our task requires our user to follow instructions or make a decision, then we can help promote a greater understanding of the task with our design inputs. We could use embedded warnings or somehow make it easier to succeed at the task or harder to fail. We can do that through simplifying or standardizing to other types of actions.

Action design inputs focus on how users act.

If our task required our users to take a particular action they didn't take, then we can look to eliminate the reason why this happened. This includes tasks that are unnecessary, unrealistic, inconvenient, or take too much time. We can look to address those problems with our design inputs. We can also apply mistake proofing to an action task.

Use a template to evaluate tasks with the PCA Framework.

While we're evaluating tasks, we want to remember the whole concept space. Our process is still within this space, and we want to be consistent with the inputs and the outputs. We're going to use the PCA framework to explore potential use errors so we can design for them.

We first think of our high-level functions. This is what we've been doing in our process flowcharts. As we develop our product, we're going to define specific tasks for our users: turn this, lift this, etc. Next, we think about the task requirements and how they relate to our user and PCA.

Since we are in concept development, we may not be talking about actually observing a customer using a product. Still, we explore the PCA framework to better understand the design inputs we need to incorporate into our product design.

Here is an example of a task analysis:

First, we break important tasks into smaller steps. For now, we think about the task as related to a function, defining what needs to be accomplished. We're empathetic to our users. What do our users need to perceive, understand, or do for this task to be successful?

Next, we can think about errors users might make at each step. We ask questions like:

- Was it an inadvertent action where they forgot something?

- Was it a thinking error, where they thought they knew they were doing the right thing, but they ended up making a mistake and doing something different?

- Or was it a deliberate deviation? And why did it occur?

From there, we can discuss the consequences if users don't accomplish this task as we expected. We can start developing design inputs to eliminate, reduce, or control the error.

Table 13.1. a Task Analysis Table

Function	Task	Task Requirement What needs to be:	Use Error	Consequence	Design Input
What function is this task related to?	What needs to be accomplished?	Perceived	• Inadvertent action • Thinking error • Deliberate deviation	What is the consequence that this task is not accomplished as we expect?	How can we eliminate, reduce, or control the error?
		Understood (Cognition)			
		Acted On			

Example Scenario: Evaluate tasks with a PCA framework.

Our scenario is that we're developing the next generation bike stand, and we conducted process flowchart exercises in which we highlighted a particular task: "Mount the bike in the stand while the bike rests on the floor."

We also make note of our motives for inspecting this function, which helps us focus on our goals during the task analysis. From our scenario, our team highlighted "Mount the bike in the stand while the bike rests on the floor" as an interface requirement (from our critical to quality process flowchart). They also highlighted it as not adding value (from our value-added flowchart). Given these two pieces of information about this use step, we want to make it as simple as possible.

Evaluate use errors.

The next step is to consider some concepts and use errors related to this task so we can pull out design inputs. Our concept design involves securing the front tire so it doesn't move. To do this, we may want to use a strap that's connected to the bicycle stand.

If we have this task as part of our concept design, what needs to be perceived, understood, and acted on for this task to be successful? Our team thinks the user needs to see and notice the tether. Our user would need to know how the tether secures the front tire. And they need to know to snap the tether closed for it to work.

Table 13.2. A task analysis table for our example scenario

Function	Task	Task Requirement What needs to be:	Use Error	Consequence	Design Input
Mount bike in stand while bike rests on the floor. *CTQ: interface* *Non-Value-Added*	Secure the front tire from moving.	P – User must see the tether.	Doesn't the tether –user forgets it (lapse)	Front wheel is not secured, and rim may get bent at the next function (raise bike in stand)	The tether shall be a color X.
		C – User must know how the tether secures the front tire.	Secures the tether to the rim instead of the frame (a mistake)		?
		A – User must snap the tether closed.	User is unable to apply enough pressure to snap the tether closed.		The force required to snap the tether closed shall be less than X N.

The next thing we want to think about is the particular use error associated with these task requirements. We break it out with PCA, like this:

- For perception, the use error is that they don't use the tether because they forget about it.

- For our cognition, the user must know how the tether secures the front tire. Our use error could be that our user secures the tether to the rim instead of the frame, which is a mistake.

- For our action requirement, our user must snap the tether closed. The use error could be the user cannot apply enough pressure to snap the tether closed.

In these cases, the front tire didn't get secured, so the rim may become bent when the user takes the step of raising the bike in the stand.

Develop design inputs.

Now we want to consider design inputs. What would be the design inputs that could address these types of use errors? If it's a perception use error, we could reduce or remove distractions. We could mistake proof or add alerts.

If it's a cognition problem, we want to make it easier to succeed or harder to make an error. We want to simplify, standardize it, or promote understanding and raise awareness of the consequences. To prevent errors during tasks, we prioritize user-friendly design and avoid taking shortcuts.

For our design input related to perception, the use error scenario is they don't see the tether or forget it. We can have a design input that the tether shall be a bright color—maybe red or neon yellow.

For our last task requirement, the action where the user must snap the tether closed, our design input could be a maximum force required to snap the tether closed. There are different standards we could reference to determine what the proper snap force could be. We can use those standards or conduct an independent study if this is a custom snap.

For the cognition task requirement, where the user must know how the tether secures the front tire, our use error scenario is they secure it to the rim of the wheel instead of the frame. We're not really sure what kind of design input we need there yet. We might have made our concept design too complicated. Or are there ways we can design the tether so our users tie it to the bike frame and not the rim of the wheel?

We're far enough along in our concept design to have tasks, but not too far that we can't consider alternatives. This would be a kind of design input option we'd want to consider as part of our design concept. This task analysis has helped us develop design inputs and prompted us to re-evaluate our understanding of user capabilities in completing this task. Not because it's our user's fault, but maybe because our product design needs to be changed.

The PCA model and a Task Analysis are just the beginning of what you can use for design development.

We evaluated high-level functions of the use process in the Concept Space. Then, we analyzed those tasks using flowcharts to identify which are critical, value-added tasks or to better understand multiple users. By prioritizing our tasks, we could confidently focus on what we could learn for design inputs using a task analysis.

The more complex our product, the higher the likelihood we'll have use errors. If our tasks and list of tasks become complex, we may want to revisit or do more task analysis to make sure we have the design inputs to address them.

There are many uses for our task analysis for other product development methods. We can use it to help us develop the product interface. Human factors engineering and usability engineering can use it for their inputs. Part of these activities is study planning, which will take place with users. As we get test results back from those studies, the PCA Model and task analysis can help us understand the root cause for any errors we noticed during the test.

Understanding the PCA Model can also help us with labeling and instructions, identifying what our users really need to know about our product. This is especially true if something particular about our product differs from other products.

Product Interface Development
Human Factors Engineering
Usability Engineering
Study Planning
Root Cause Analysis
Task Analysis ➡️ Labeling and Instructions
User Training Programs
Risk Analysis (UFMEA)
Reliability
Quality

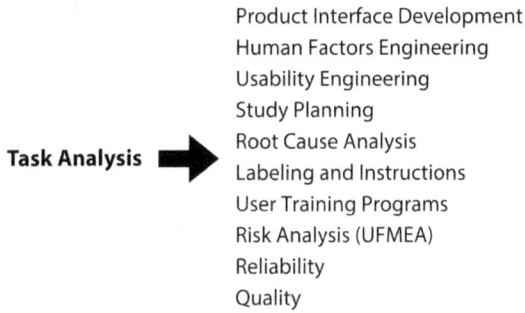

Figure 13.3. Uses for a Task Analysis

We can also use the PCA Model for risk analysis in a system FMEA or a Use FMEA. We can take the task analysis and put it into an FMEA to evaluate the severity and the occurrence, gauge the risk, and prioritize design activities.

Reliability engineers can use the model to help them develop test plans and better understand the limits of a design. Quality engineers can use it to determine what features need to be monitored and how that links to usability.

Starting concept development with a task analysis is a good idea, and you can share it with the team members who will use it throughout development.

Because a task analysis like this can get pretty large and unwieldy, I recommend choosing only those parts of the process that are critical. Use the Concept Space Model and the Use Process analyses to identify critical areas before you start a task analysis.

Again, we want to move forward with the design, but we want to do it with information. A task analysis can help us better understand our users and any behaviors that may lead to errors. And that gives us design inputs.

Key Takeaways:

1. A use error is a deviation from what is expected to happen during product use, and the use process itself is what needs to be examined for errors, not the user. The three types of use errors are inadvertent action, thinking error, and deliberate deviation.

2. The PCA (Perception, Cognition, Action) framework is a model of user interaction with a product that involves how users perceive the output of a product, think about it and process information, and then decide how to act. The PCA framework can be used to analyze tasks and identify where errors may occur.

3. Design inputs can be developed to eliminate or control use errors related to perception (reducing distractions or mistake proofing), cognition (promoting understanding of the task), and action (eliminating unnecessary or inconvenient tasks).

4. Task analysis helps break down complex processes into smaller steps and determine what users need to perceive, understand, and do for successful task completion.

5. Task analysis can be used in other product development methods, such as developing the product interface, planning usability studies, creating labels and instructions, and risk analysis.

Reflection Questions:

1. In a recent project, how did I treat the use error? Did I blame the user, or did I consider how the design could change the use process?

   _____

   _____

2. When I hear about a use error, what is the immediate next step I should take toward understanding the error?

   _____

   _____

3. Where could I look to identify potential use errors for concept development beyond the design team, such as regulatory, marketing, and customer service groups?

   _____

   _____

Part Three Review

Key Concepts

1. Early concept development should inform later design phases, using tools like the Concept Space Model and iterating the initial ideas throughout the product development process.

2. Drivers to benefits are used to develop design inputs by mapping vague drivers to specific design inputs using tree diagrams and then using matrices to compare these with customer satisfaction.

3. Analyzing customer symptoms involves breaking them down into outcomes and impacts, assigning probabilities, and using this information to prioritize design actions and inform an FMEA for risk analysis.

4. Use errors should be analyzed using the PCA framework to find causes of errors (perception, cognition, or action) and then developing design inputs to eliminate or control them.

5. The ADEPT Team Framework is adaptable. It can continue to be used with visual models and templates to facilitate team sessions to share knowledge, reach consensus, and generate design inputs.

Practice Activities

1. List at least two ways you can ensure the valuable insights and information gathered during early concept development are effectively carried through to the later stages of product design.

2. **Develop a Tree Diagram and Matrix:** Follow these steps to practice using the Benefit-Impact Template to understand features

and impacts and how to prioritize early design activities and define design inputs.

2a. Choose a product or feature you are currently working on or one you would like to improve.

2b. Use the Benefit-Impact Template to identify the drivers for a specific benefit you aim to achieve.

2c. Create a tree diagram to break down these drivers into specific, actionable design inputs.

2d. Use an L-shaped matrix to compare your features to the design inputs you have developed. Then use a roof-shaped matrix to explore how the design inputs correlate or compete with one another.

2e. Evaluate these relationships to determine which inputs are most important for quality, reliability, and cost.

3. **Conduct a Symptom Risk Analysis and start an FMEA:** Follow these steps to practice using the Symptom-Impact Template to understand outcomes and impacts and how to prioritize early design activities and define design inputs using these analyses.

3a. Think of a product you use frequently and identify a symptom you or another user might experience while using the product.

3b. Break this symptom down into an outcome and an impact.

3c. Assign probabilities to both the outcome and the impact and use these to calculate a risk priority number.

3d. Create a simple risk map using the severity of the impact and the likelihood of the symptom.

3e. Take the information you gathered and input it into a simple FMEA table. Preparing for FMEA helps you move beyond identifying symptoms to developing concrete, risk-based design inputs.

4. **Perform a Task Analysis Using the PCA Framework:** This exercise helps you understand how use errors relate to the process itself rather than the user and will guide you in designing more user-friendly products or processes. This will also give you practice in applying the PCA framework and developing design inputs that focus on user experience.

   4a. Select a specific task related to a product or service you want to analyze.

   4b. Using the PCA (Perception-Cognition-Action) framework, break down what a user needs to perceive, understand, and do to complete the task successfully.

   4c. Identify potential use errors at each step.

   4d. Develop design inputs to eliminate, reduce, or control these errors.

Conclusion

BUILDING UPON STRONG CONCEPTS

You are now not only poised to recognize the challenges in your team's concept development, but you also have ways to improve it. You have frameworks and ideas to help you and your team develop design inputs before diving into engineering solutions.

The methods and techniques described in this book are tools for your team to engage in better product design and development to create products that truly meet your users' needs and expectations.

Your goal was to engineer a solution your customer wanted and avoid the disappointment of engineering one that was not. Now, you have an effective process that is focused and does not take a lot of time. By using a repeatable method, frameworks for discussion, and methods to prioritize potential customer experiences, you are now equipped to make better design decisions.

The models and templates provided helps you organize your ideas, create context, and break down complex problems into manageable parts. By focusing on co-working with a cross-functional team, you can move past the difficulties that often arise during concept development. By following a consistent approach during meetings, you can ensure the team stays on track and works towards common goals.

By now, you see how concept development links to design inputs, user needs, and other design constraints. The ultimate success of concept development is design information that your team can develop into design inputs.

It's important to remember that concept development is not the end but the beginning of a vital conversation. The work you do in this phase will serve as a foundation for the entire product development process.

You've broken down ideas to translate them into design inputs. During the design process, you're building them back and putting them together into a product. You must step back and look at the bigger picture during design.

"You think because you understand *one* you must also understand *two*, because one and one make two. But you must also understand *and*." -Attributed to Rumi, 13th century Persian poet

Rumi's quote reminds us that interfaces and the bigger picture don't always make sense when we keep drilling down into the details. Use the Concept Space Model throughout development to check that the whole design is coming together to meet the parts we explored in concept development. Rely again on your team to compare your proposed solution against the concept space model. Does it fit or have we deviated? Have we learned new information that leads us to change our mind?

The methods you learned are not just for concept development—they can be easily adapted to other team activities. You can use the ADEPT Team Framework for analysis and design decisions throughout product development. Recognizing that models and templates help teams share knowledge, you can develop your own for any areas of design where you need to engage teamwork.

Next Steps

Remember, you do not need to implement all these ideas all at once. To continue your journey, consider these next steps based on the methods described in this book:

Review your current projects: Identify opportunities to apply the ADEPT Team Framework and Concept Space Models to ongoing or upcoming projects.

Schedule team sessions: Plan co-working sessions with your cross-functional teams, using the models and templates discussed in the book. Remember to set clear goals for each meeting.

Start with alignment: Begin each co-working session by aligning your team around the problem space. Make sure everyone understands the users, the use space, and the needs your new product is trying to fill.

Use visual aids: To help your team understand complex ideas, use visual aids like process flowcharts, models, and templates. This will allow you to break down large concepts into smaller, more digestible pieces for knowledge sharing. Customizing models and templates to your specific needs helps ensure you get better ideas.

Prioritize user experiences: Always keep your focus on user experiences in concept development: benefits, symptoms, and use processes. Evaluate features using methods such as the Kano Model or severity ratings.

Link drivers to design inputs: Use the methods in this book to connect the drivers of benefits and symptoms directly to design inputs.

Integrate into the design process: Don't let the outputs of concept development sit dormant. Use the information gathered in early concept development to inform later phases of development, such as FMEA, task analysis, and risk assessments.

Follow-up and follow-through: Show your team how their contributions helped move the design forward.

Adapt the approach: Use the models, templates, and ADEPT Team Framework to conduct other activities with your team.

Continue to learn: Keep an open mind and view each co-working session as an opportunity to learn more about your users and your product design.

By implementing these steps, you and your team can successfully navigate the concept space, develop innovative products, and achieve your goals.

ACKNOWLEDGEMENTS

Thank you to my quality and reliability engineering mentors: Jeffrey, Ronito, and David A. You sparked my interest and then really fueled it.

I appreciate the wonderful people with whom I've had the chance to work. They patiently trusted me with new ideas, shared their knowledge, and provided valuable feedback. I consider myself fortunate to have collaborated with professional, results-driven individuals who genuinely care about the customers they serve.

My family is my daily support system.

Thank you to my parents, Bill and Carol, for not isolating me from entrepreneur life and allowing me to learn from doing. You showed me the power of dedication and encouraged my decisions. I'm also grateful for the consistent support of my parents and my older sister, Brenda, who's an engineer herself. Love to you all!

To my husband, Vincent: I am so very grateful for your support in all things, including this book and my business. I can bury myself in work, but thanks to you, I'm learning to find a better balance. No one could be a better husband, best friend, traveling companion, and hobby enthusiast partner than you. I love you!

To L: Thank you for showing me how to use new technology and being a source of confidence for me to finish this book. Love you.

To A: Thanks to you, I learned to write better. I'm glad we had a dialogue about that. Love you.

REFERENCES/NOTES

Introduction

Cooper, Robert G. *Winning at New Products: Accelerating the Process from Idea to Launch,* 3rd ed. Basic Books, 2001.

Chapter One

"Eleven Lessons: Managing Design in Eleven Global Brands." Design Council. Accessed March 20, 2025. https://www. designcouncil.org.uk/fileadmin/uploads/dc/Documents/ ElevenLessons Design Council%2520%25282%2529.pdf.

Jokela, Timo, Netta Iivari, Juha Matero, and Minna Karukka. "The Standard of User-Centered Design and the Standard Definition of Usability: Analyzing ISO 13407 Against ISO 9241-11." *CLIHC '03: Proceedings of the Latin American Conference on Human-Computer Interaction* (2003): 53–60, https://doi.org/10.1145/944519.944525.

Reinertsen, Donald, and Stefan Thomke. "Six Myths of Product Development." *Harvard Business Review.* May 2012. https://hbr. org/2012/05/six-myths-of-product-development.

Chapter Two

Lindholm, Christin, and Martin Host. "Risk Identification by Physicians and Developers: Differences Investigated in a Controlled Experiment." *2009 ICSE Workshop on Software Engineering in Health Care* (2009): 53–61, https://doi.org/10.1109/SEHC.2009.5069606.

Dale Carnegie. *How to Win Friends and Influence People: Updated for the Next Generation of Leaders.* Simon & Schuster, 2022.

Stanier, Michael Bungay. *The Advice Trap: Be Humble, Stay Curious and Change the Way You Lead Forever.* Box of Crayons Press, 2020.

Newport, Cal. *Deep Work: Rules for Focused Success in a Distracted World.* Grand Central Publishing.

Schwarzenegger, Arnold. *Be Useful: Seven Tools for Life.* Penguin Press, 2023.

Chapter Four

Boyd, Drew, and Jacob Goldenberg. *Inside the Box: A Proven System of Creativity and Breakthrough Results.* Simon & Schuster, 2013.

Eppler, Martin J., Heidi Forbes Öste, and Sabrina Bresciani. "An Experimental Evaluation on the Impact of Visual Facilitation Modes on Idea Generation in Teams." 17th International Conference on Information Visualisation, London, UK, 2013. https://doi.org/10.1109/IV.2013.43.

Perez, Marta, and Sabrina Brescaiani. "The Role of Visual Templates on Improving Teamwork Performance." *19th International Conference on Information Visualization*, 2015. http://dx.doi.org/10.1109/iV.2015.66.

Chapter Five

Boyd, Drew, and Jacob Goldenberg. *Inside the Box: A Proven System of Creativity and Breakthrough Results.* Simon & Schuster, 2013.

Knapp, Jake, John Zeratsky, and Braden Kowitz. *Sprint: Solve Big Problems and Test New Ideas in Just Five Days.* Simon & Schuster, 2016.

De Dreu, Carsten K. W., Bernard A. Nijstad, Matthijs Baas, Inge Wolsniok, and Marieke Roskes. "Working Memory Benefits Creative Insight, Musical Improvisation, and Original Ideation Through Maintained Task-Focused Attention." Personality and Social Psychology Bulletin, 38, no 5. 2012. https://doi.org/10.1177/0146167211435795.

Schmitt, Lara, Stéphanie Buisine, Jonathan Chaboissier, Améziane Aoussat, Frédéric Vernier. "Dynamic Tabletop Interfaces for Increasing Creativity." *Computers in Human Behavior* 28, no. 5. 2012. https://doi.org/10.1016/j.chb.2012.05.007.

Chapter Six

"Brain Writing: Lean Six Sigma." Six Sigma Certification. Accessed March 20, 2025. https://www.sixsigmacertificationcourse.com/glossary/brain-writing.

Appendix A
PROMPTS EXPLORING POSITIVE VERSUS NEGATIVE SYSTEM OUTPUTS

I entered two different prompts within Gemini with 1.5 Flash AI. I wanted to illustrate how viewing a problem positively (achieving benefits) differs from viewing it negatively (avoiding symptoms). It's worth considering both perspectives, and the prompts generated different responses.

Prompt 1

I first focused on a negative experience, using the Symptom-Impact statement format.

Prompt: What are potential design features and service options to avoid this customer experience? [Our customers] may [not be able to fit their bike to the rack] which leads to [frustration].

Output 1

AI generated this analysis:

Design Features and Service Options to Avoid
Customer Frustration

Understanding the Problem: It seems your customers are experiencing difficulties fitting their bikes onto the bike rack. This could be due to various factors such as bike size, rack design, or lack of clear instructions.

Potential Solutions

Design Features:

1. **Adjustable Rack**

   - **Height:** Allow the rack to adjust to different bike sizes.

   - **Wheelbase:** Consider a design that accommodates various wheelbases.

   - **Angle:** Make the rack adjustable to fit different bike angles.

2. **Universal Compatibility**

   - **Wheel Size:** Ensure the rack can accommodate various wheel sizes, including mountain bikes and road bikes.

   - **Frame Types:** Design the rack to be compatible with different frame types (e.g., step-through, traditional).

3. **Clear Labeling and Instructions**

   - **Step-by-Step Guide:** Provide clear and easy-to-follow instructions, possibly with diagrams or illustrations.

   - **Labeling:** Label different parts of the rack for easy identification and use.

4. **Safety Features**

- **Locking Mechanisms:** Ensure the rack has secure locking mechanisms to prevent bikes from falling off.

- **Weight Capacity:** Clearly indicate the maximum weight capacity to avoid overloading.

Service Options:

1. **Customer Assistance**

- **On-Site Assistance:** Provide staff to assist customers with fitting their bikes to the rack.

- **Online Tutorials:** Offer video tutorials or online guides to help customers.

2. **Bike-Fitting Service**

- **Professional Assistance:** Offer a bike-fitting service where customers can bring their bikes to a designated location for assistance.

3. **Return Policy**

- **Hassle-Free Returns:** Have a generous return policy for customers who find the rack unsuitable for their needs.

4. **Feedback Mechanism**

- **Customer Surveys:** Gather feedback from customers to identify areas for improvement and address their concerns.

By implementing these design features and service options, you can significantly reduce customer frustration and ensure a positive experience.

Prompt 2

Next, I changed the prompt into a positively worded customer experience, using a Benefit-Impact Statement format.

Prompt: What are potential design features and service options to meet this customer experience? [Our customers] need [to fit their bike to the rack] so they can [feel satisfied].

Output 2

AI came up with this analysis:

Design Features and Service Options to Enhance Customer Satisfaction

Understanding the Need: Your customers want to be able to fit their bikes to the rack easily and efficiently. This sense of accomplishment and control contributes to their overall satisfaction.

Potential Solutions

Design Features:

1. **Intuitive Design**

    - **Clear and Simple:** Make the rack's design intuitive and easy to understand, even for those who are not mechanically inclined.

    - **Visual Cues:** Use visual cues, such as color coding or symbols, to guide users through the process.

2. **Easy-to-Use Mechanisms**

    - **Quick-Release Features:** Incorporate quick-release mechanisms or levers to make it easy to adjust and secure the bike.

    - **Smooth Operation:** Ensure that all moving parts operate smoothly and without excessive force.

3. **Secure Fit**

    - **Locking Mechanisms:** Provide reliable locking mechanisms to ensure the bike is securely attached to the rack.

- **Adjustable Straps:** Include adjustable straps or clamps to accommodate different bike sizes and styles.

4. **Durable Construction**

   - **High-Quality Materials:** Use durable materials that can withstand the weight of bikes and outdoor elements.

   - **Weather Resistance:** Design the rack to be resistant to rust, corrosion, and other environmental factors.

Service Options:

1. **Educational Resources**

   - **User Manuals:** Provide clear and concise user manuals with step-by-step instructions.

   - **Online Tutorials:** Offer online tutorials or videos demonstrating how to use the rack.

2. **Customer Support**

   - **Responsive Support:** Provide responsive customer support channels, such as phone, email, or live chat, to address any questions or concerns.

   - **Troubleshooting Guides:** Offer troubleshooting guides to help customers resolve common issues.

3. **Community Engagement**

   - **Social Media:** Encourage customers to share their experiences and tips on social media.

   - **Online Forums:** Create online forums or communities where customers can discuss the rack and offer advice.

4. **Rewards Program**

   - **Loyalty Incentives:** Implement a rewards program to incentivize customers to use the rack and provide feedback.

By incorporating these design features and service options, you can create a positive customer experience that fosters satisfaction and loyalty.

Appendix B
COMPARISON
FLOWCHART ANALYSIS

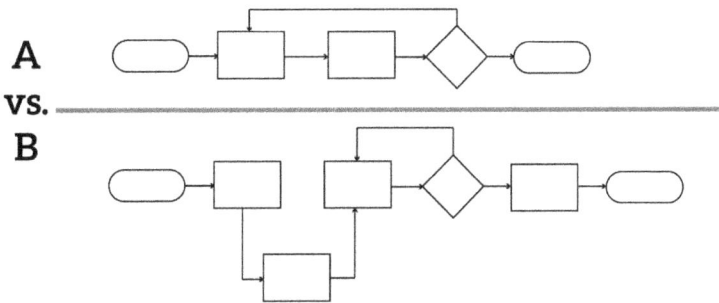

Figure 10.3.

Start with two processes: the use process you drew for the Concept Space Model and another target process. At the end of this exercise, you'll have highlighted areas of the use process that need special consideration and focus during design. These areas helps you compete or improve upon other options.

We need to do multiple Discover prompts for this exercise. Follow these steps to use the ADEPT Team Framework for a Comparison Flowchart exercise:

Do pre-work. Gather customer descriptions and preliminary needs. Plus gather flowcharts that you'll be reviewing. Draw two flowcharts, A and B, or edit one flowchart and highlight the changes. Review the basic flowchart shapes. Review the ADEPT Team Framework to get ready to lead the meeting.

1. **Align** on a goal and scope of the working meeting.

    1a. Review or make available the flowcharts A and B that you'll compare or the flowchart you'll analyze.

    1b. Make the goal visible on agendas and during teamwork. An example: "Highlight differences between our concept product and competitor X and what we want to prioritize for design of our product."

    1c. Agree on a scope. Describe the flowcharts you're reviewing. An example: "Compare our concept use process with competitor X, which starts at [start point of flowchart] and ends at [end point of flowchart]."

2. **Discover** design ideas from the team.

    2a. Each team member brain writes based on prompts and shares after each prompt. Consider B the desired, future state.

    - First prompt: What IS NOT working? On pink Post-its, write what IS NOT working in B for any process or decision step. What do we want to do differently in

A? Share the ideas by placing the pink Post-its near the step in B.

- Second prompt: What IS working? On green Post-its, write what IS working in B for any process or decision step. What do we want to carry-over into A? Share the ideas by placing the green Post-its near the step in B.

- Third prompt: What is new? On blue Post-its, write what we want to do that is not in A or B. What do we need to learn to implement this properly? Share the ideas by placing blue Post-its near the step in A.

2b. All team members have shared their ideas, placing them near the process step or decision for all to read.

3. **Examine** the ideas for clarity and understanding. A facilitator summarizes the group's findings for each process step, both what is and is not working, for understanding and clarity. They also summarize the additions using blue Post-its. This expands the group's understanding of the desired-state use process.

4. **Prioritize** changes. Rank the proposed changes for importance for design implementation.

4a. Multi-voting: Display the process flowcharts and results. Each team member gets five dots (sticker or drawn) to place on the flowchart, corresponding to what they believe is important to implement.

4b. Review and discuss the results. Aim for consensus, which is that place where everyone supports the decision even if it wasn't their first choice. If there is no consensus, consider going back to the Examine step to ensure all team members are voting on the same idea or the same understanding of an idea.

5. Perform the **Teamwork** activities to close the meeting. List action items for follow-up, including parking lot items. Take pictures or save whiteboards.

6. Design for the use process. Summarize this team design activity with minutes and share what was learned. Use teamwork results to link needs with requirements.

    6a. Summarize important functional steps and use decisions and share with the team.

    6b. Update user descriptions and their needs, if necessary.

    6c. Prioritize implementation of use steps through features and offerings.

After a Comparison Flowchart analysis, you'll better understand differences and have more information against which to design.

Appendix C
CRITICAL TO QUALITY ANALYSIS

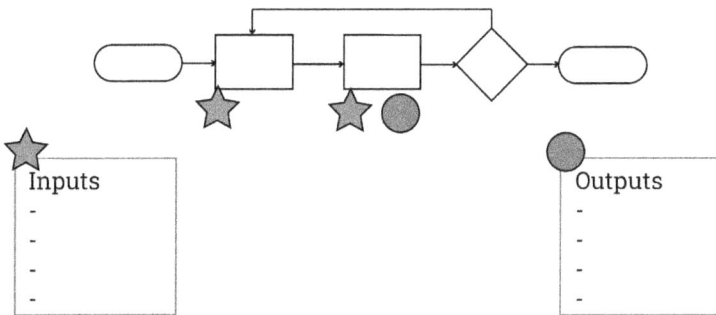

Figure 10.4.

The basic steps of performing a Critical to Quality analysis involve asking a series of questions:

1. What are more specific outputs? Identify any hidden customer needs or benefits.

2. What process steps affect output? Identify functional priorities. Mark those steps with a label or symbol. In our example, we use a dot.

3. What are more specific inputs? Clarify users, use environment, and other conditions and assumptions of our users. We meet customers where they are.

4. What steps are affected by inputs? Identify interface requirements. Mark those steps with a label or symbol. In our example, we use a star.

Start with a flowchart, either the use process from the Concept Space Model or from an Alignment Flowchart analysis. When you are done with a Critical to Quality analysis, you'll have clarified your users at the input and output. You'll also have identified process steps that are critical to quality, both for achieving desired outputs and that are sensitive to inputs.

We need to do multiple Discover prompts for this exercise. Perform these ADEPT Steps to help you plan and facilitate your team through a Critical to Quality Analysis:

Do pre-work. Gather customer descriptions, preliminary needs, and a flowchart. Review the basic flowchart shapes. Review the ADEPT Team Framework to get ready to lead the meeting.

1. **Align** on a goal and scope of the working meeting.

    1a. Review the flowchart you'll analyze.

    1b. Make the goal visible on agendas and during teamwork. An example: "Identify which use process steps are critical to quality."

    1c. Agree on a scope. Describe the flowchart you're reviewing. An example: "We're reviewing a use process for [customer], which starts at [start point of flowchart] and ends at [end point of flowchart]."

    1d. Summarize the inputs and desired outputs your team has developed so far, focused on the user's state. List the conditions and assumptions of the users at the input and the desired benefits at the output.

2. **Discover** design ideas from your team. Each team member brain writes based on prompts and shares after each prompt, placing their idea next to either the input or output.

    2a. First prompt: What are more specific outputs? Identify any hidden customer needs or benefits.

    2b. Second prompt: What are more specific inputs? Clarify users, use environment, and other conditions and assumptions of our users. We meet customers where they are.

3. **Examine** the ideas for clarity and understanding. A facilitator summarizes the group's findings at the outputs and inputs. This expands the group's understanding of the inputs and outputs.

4. **Prioritize** process steps. Relate the proposed changes for importance for design implementation based on what affects outputs and what inputs affect steps.

    4a. Multi-voting: Each team member gets dots (sticker or drawn) to place on the flowchart.

       - What use process steps affect the output? They place a blue mark on the process steps that affect output.

       - What use process steps are affected by inputs? Place a red mark on the process steps that are affected by inputs.

    4b. Review and discuss the results. Aim for consensus, which is that place where everyone supports the decision even if it wasn't their first choice. If there is no consensus, consider going back to the Examine step to ensure all team members are voting on the same idea or the same understanding of an idea.

5. Perform the **Teamwork** activities to close the meeting. List action items for follow-up, including parking lot items. Take pictures or save whiteboards.

6. Design for use step priorities. Summarize this team design activity with minutes and share what was learned. Use teamwork results to link needs with requirements.

   6a. Design for functional priorities. Process steps marked as affecting the output are critical to quality. If performed well, they affect the quality of the output.

   6b. Design for interface requirements. Process steps marked as affected by the input are impacted by the quality of those inputs. Determine how to better control the inputs or make that process step more robust to changing circumstances.

   6c. Update user descriptions and their needs, if necessary.

After a Critical to Quality analysis, you'll better understand the functional priorities and interface requirements, both of which are important design inputs.

Appendix D
VALUE-ADDED ANALYSIS

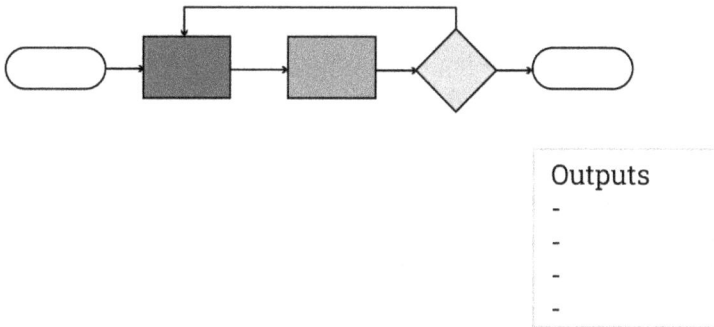

Outputs
-
-
-
-

Figure 10.6.

There are three levels of value we identify for three different actions:

1. What adds value? These are high-priority steps we want to ensure we implement and implement well.

2. What is not value-added? These are areas we want to simplify by reducing or eliminating.

3. What is something that just must be done? Perhaps it's not adding value, but we need to do it anyway. These are steps we want to make easier by removing hurdles to their completion.

Start with a flowchart, either the use process from the Concept Space Model or from an Alignment Flowchart analysis. You can also start from a Critical to Quality analysis, which your team has already worked on to define Outputs.

At the end of this exercise, the team will have prioritized use steps that are value-added, not value-added, or necessary steps. This helps you with design inputs so you better understand what to simplify, exaggerate, or eliminate.

In this analysis, we're going to use the Prioritize step to identify what is value-added (or not) using multi-voting. Follow these steps to use the ADEPT Team Framework for an Alignment Flowchart exercise:

Do pre-work. Gather customer descriptions, preliminary needs, and a flowchart. Review the basic flowchart shapes. Review the ADEPT Team Framework to get ready to lead the meeting.

1. **Align** on a goal and scope of the working meeting.

    1a. Review the flowchart that you'll analyze.

    1b. Make the goal visible on agendas and during teamwork. An example: "Identify which use process steps are value-added."

    1c. Agree on a scope. Describe the flowchart you're reviewing. An example: "We're reviewing a use process for [customer], which starts at [start point of flowchart] and ends at [end point of flowchart]."

    1d. Summarize the desired outputs that your team has developed so far, focused on the user's state. List the desired benefits at the output.

2. **Discover** design ideas from your team.

   2a. Ask, "What are more specific outputs?" Each team member brain writes based on prompts and shares after each prompt. All team members share their ideas, placing them near the output for all to read.

   2b. Or continue working from the team's outputs that they listed in a previous flowchart analysis exercise, like the Critical to Quality analysis.

3. **Examine** the output and steps for clarity and understanding.

4. **Prioritize** changes. Relate the proposed changes for importance for design implementation.

   4a. Multi-voting: Each team member gets three different colored dots (sticker or drawn) to place on the flowchart.

   - Add a green mark to steps that add value.

   - Add a red mark to steps that don't add value.

   - Add a yellow mark to steps that need to be done, but don't contribute to the quality of the output.

   4b. Review and discuss the results. Aim for consensus, which is that place where everyone supports the decision even if it wasn't their first choice. If there is no consensus, consider going back to the Examine step to ensure all team members are voting on the same idea or the same understanding of an idea.

5. Perform the **Teamwork** activities to close the meeting. List action items for follow-up, including parking lot items. Take pictures or save whiteboards.

6. Design for use steps based on the value they bring. Summarize this team design activity with minutes and share what was learned. Use teamwork results to link needs with requirements.

6a. Summarize the value-added steps, the steps that don't add value, and the necessary steps that don't add value. Share with the team.

6b. Update user descriptions and their needs, if necessary.

6c. Prioritize implementation of use steps through features and offerings.

At the end, you'll better understand what adds value to the customer and what doesn't. Prioritize value-added steps to deliver benefits to customers in the best way possible. For non-value-added steps, we reduce the effort or eliminate the step, if we can. We also make organizational needs as easy as possible.

Appendix E
DEPLOYMENT FLOWCHART

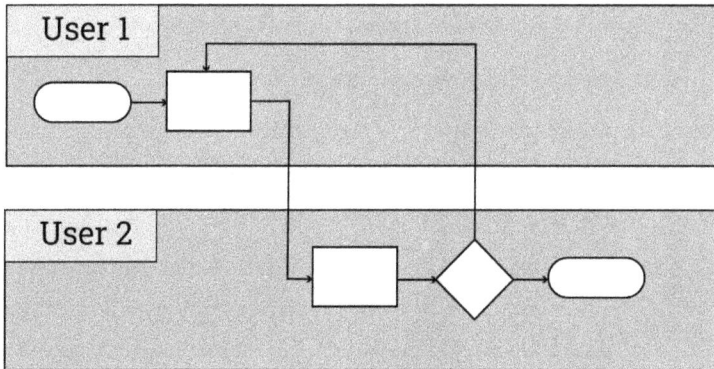

Figure 10.8.

Start with a flowchart, either the use process from the Concept Space Model or another flowchart analysis. At the end of this analysis, we'll better understand the interface between users and our product. We'll also be able to coordinate use steps among multiple users, understand parallel activities, and ensure the proper transfer of steps between users.

In this analysis, we're going to use the Discover step to identify what steps go into which swim lanes and brain write new steps that might be needed. In the Prioritize step, we'll ask the team to come to a consensus. Follow these instructions to use the ADEPT Team Framework for a Deployment Flowchart exercise:

Do pre-work. Gather customer descriptions, assemble preliminary needs, and choose a flowchart. Review the basic flowchart shapes. Review the ADEPT Team Framework to get ready to lead the meeting.

1. **Align** on a goal and scope of the working meeting.

    1a. Review the flowchart and customers you'll analyze.

    1b. Make the goal visible on agendas and during teamwork. An example: "Identify use process steps taken by customer X and those taken by customer Y."

    1c. Agree on a scope. Describe the flowchart you're reviewing. An example: "We're reviewing a use process for [customers X and Y], which starts at [start point of flowchart] and ends at [end point of flowchart]."

    1d. Summarize the inputs and desired outputs your team has developed so far, focused on the user's state. List the conditions and assumptions of the users at the input and the desired benefits at the output.

2. **Discover** design ideas from your team.

    2a. Treat this activity like an affinity diagram. Ask your team to group and move process steps to different customer swim lanes.

2b. Team members brain write additional steps that might be needed. They then share their ideas, placing this on the flowchart for all to read.

3. **Examine** the ideas for clarity and understanding. A facilitator summarizes the group's findings for each swim lane for understanding and clarity. They also summarize the additions. This expands the group's understanding of the desired-state use process.

4. **Prioritize** changes. Review and discuss the results. Aim for consensus, which is that place where everyone supports the decision even if it wasn't their first choice. If there is no consensus, consider going back to the Examine step to ensure all team members are voting on the same idea or the same understanding of an idea.

5. Perform the **Teamwork** activities to close the meeting. List action items for follow-up, including parking lot items. Take pictures or save whiteboards.

6. Design for benefits. Summarize this team design activity with minutes and share what was learned. Use teamwork results to link needs with requirements.

   6a. Summarize important functional steps and use decisions and share with the team.

   6b. Update user descriptions and their needs, if necessary.

   6c. Prioritize implementation of use steps through features and offerings.

At the end of the activity, the team will better understand who is doing what and when. We may need to revisit aspects of the Concept Space due to multiple users. This includes the use process inputs, outputs, and the Concept Space Model.

Appendix F

CHECKLIST FOR EVALUATING A MATRIX DIAGRAM

Explore the relationships and correlations of your matrix diagram and label their relationship in the relationship matrix and correlation matrix. Use the following checklist to help you think about the information you developed with your team in the L-shaped and roof-shaped matrix. Then decide how to change your design inputs.

- Ensure our features aren't being duplicated.

  o No identical rows.

- Ensure we have technical design inputs that enable the features that matter.

  o No empty rows.

  o Each row has one strong relationship to a characteristic.

- Prioritize our list of design inputs.

  o Which inputs are associated with specific levels of customer satisfaction?

  o What is the strength of that association?

- Understand how design inputs are linked.

  o Which compete for trade-offs?

  o Which ones support more than one need?

A successful matrix analysis looks like:

- List of design inputs that are linked to benefits/features and their customer satisfaction rating

- Understanding of what to change about the developing concept features and design inputs to best meet the needs of the concept space

- Action items to research:

  o Concept space to better understand relationships

  o Design inputs to understand their relationship to features and to each other

The team can reuse this matrix diagram to help them make design decisions later in the product development process.

Appendix G
CONDITIONAL PROBABILITIES

The Symptom-Impact Model asks us to estimate the likelihood of the Outcome occurring and the likelihood of the Impact given that the Outcome has occurred. It is easier to think of an event in this way because it narrows our focus, and we can use the drivers to help us estimate these likelihoods.

We may consider those values as estimates of probability. We then may multiply these two probabilities to estimate the overall likelihood of both the Outcome and Impact.

We can prioritize our ideas with any one of these three likelihood values.

For a better understanding of conditional probabilities (what we are using), this sheet outlines their mathematical rules. You do not need to explain this in your team meetings.

Remember that we are estimating based on our current knowledge or assessing what we don't know and where we need to do some investigations. For example, if we find we have an event with a high severity and a high overall likelihood of occurrence, we may want to plan to investigate or conduct tests to learn more about the drivers and better understand the actual occurrence values.

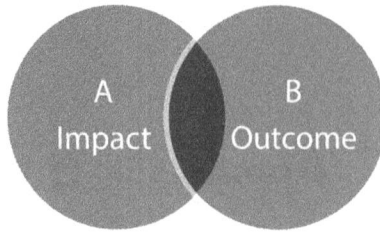

Figure G.1. Conditional Probabilities

P(B) and P(A|B): This is what we ask ourselves to estimate in the Symptom Break-Down Model for the Outcome and Impact, respectively.

We multiply these to estimate P(A ∩ B) for a risk event.

P(A) = Probability of event A, Probability that the IMPACT has occurred.

*P(B) = Probability of event B, Probability that the OUTCOME has occurred.

*P(A ∩ B) = Probability of event A and event B occurring, the probability of the Outcome and Impact occurring. This is the intersection of these two events, where the Venn diagram overlaps.

*P(A|B) = Probability that event A has occurred given that event B has occurred, Probability that the Impact has occurred given that the Outcome has occurred.

Mathematically, this value is a ratio of the intersection and an original probability.

$$P(A|B) = P(A \cap B) / P(B)$$

This is a conditional probability. The reason P(B) is in the denominator is that if event B is known to have occurred (the Outcome), then the sample space is reduced to just event B.

The overall likelihood of the outcome and impact occurring, or the occurrence of the risk event:

$$P(A) \text{ and } P(B) \rightarrow P(A \cap B) = P(A|B) \times P(B), \text{ provided } P(B) > 0$$

Appendix H
OTHER RECOMMENDED RESOURCES

Choosing a new product project:

Cooper, Robert G. *Winning at New Products, Third Ed.* Basic Books. 2001.

Merrill, Peter. *Innovation Never Stops: Innovation Generation – the Culture, Process, and Strategy.* United States, ASQ Quality Press, 2015.

Phillips, Jack J., and Phillips, Patricia Pulliam. *The Value of Innovation: Knowing, Proving, and Showing the Value of Innovation and Creativity.* Germany, Wiley, 2017.

Wunker, Stephen, et al. *Jobs to Be Done: A Roadmap for Customer-Centered Innovation.* United States, AMACOM, 2016.

Improving your facilitation skills:

Grenny, Joseph, et al. *Crucial Conversations: Tools for Talking When Stakes are High, Third Edition.* United States, McGraw Hill LLC, 2021.

Kaner, Sam. *Facilitator's Guide to Participatory Decision-Making.* Germany, Wiley, 2011.

Learning more about cross-functional teams:

Finerty, Susan Z. *Master the Matrix: 7 Essentials for Getting Things Done in Complex Organizations.* United States, Salem Author Services, 2012.

Parker, Glenn M. *Cross- Functional Teams: Working with Allies, Enemies, and Other Strangers, Completely Revised and Updated.* Germany, John Wiley & Sons, 2015.

GLOSSARY

ADEPT Team Framework: A process for team collaboration during concept development. The framework includes the steps of Align, Discover, Examine, Prioritize, and Teamwork. As both a planning and execution process, it helps teams share knowledge and prioritize against goals and next steps.

Affinity Diagram: A method used to group and organize ideas, often using Post-it notes, to identify subheadings and themes.

Benefits: The positive outcomes or impacts a product has on a user. They describe the user's experience and are linked to product features.

Brain Writing: A technique for generating ideas where team members write, draw, or type their ideas individually in response to a prompt. This is done silently in the same physical or virtual space within a set time limit. It's used as an alternative to traditional brainstorming to avoid issues like dominant personalities, groupthink, and evaluation apprehension. By allowing everyone to contribute independently and anonymously, it aims to increase the quantity and diversity of ideas. The collection of ideas can then be used as a basis for further examination, grouping (e.g., using affinity diagrams), and prioritization in the concept development process.

Concept Development: An early stage of new product development, during which teams work together to share knowledge and develop ideas before designing solutions. It involves understanding the use space and customer expectations. The goal is to gather information for design inputs.

Concept Space Model: A tool for team discussion and knowledge sharing during concept development. It helps teams explore user experiences through the focus areas of benefits, symptoms, and use processes. It includes the boundaries of the use space and customer expectations of a new product.

Conditional Probability: The probability of an event occurring based on another event having already occurred, which can be used to assess risk.

Consensus: A state when the team agrees on a decision, even if it wasn't their top choice. It doesn't involve forcing team members to switch their votes. The aim is for everyone to support the final decision, even if they initially favored another option.

Correlation Matrix: Part of the matrix diagram used to identify relationships between design inputs.

Critical to Quality (CTQ): Features associated with drivers to must-have design inputs (refer to the Kano Model categories).

Critical to Quality Analysis: A type of use process analysis that identifies functional priorities and interface requirements important for quality.

Critical to Motivation (CTM): Features associated with drivers to attractive design inputs (refer to the Kano Model categories).

Critical to Satisfaction (CTS): Features associated with drivers to one-dimensional design inputs (refer to the Kano Model categories).

Cross-functional Team: A team composed of people from different departments or areas of an organization working together on a project. These teams bring different perspectives to create a better product.

Customer Experience: A customer's overall impression (either positive or negative) of a product based on their usage.

Design Inputs: Information used as guardrails for designing a solution. Examples include requirements, risk controls, performance criteria, user interfaces and experience, environmental conditions, and standards.

Design Outputs: The result of the design process, such as technical specifications, drawings, 3D models, and prototypes.

Deployment Flowchart: An analysis of a process flowchart to understand who is doing what and when, especially when there are multiple users.

Directed Co-work: Facilitated meetings with a team to work toward a common, focused goal.

Drivers: Ideas that make a feature or outcome possible or increase their impact. They are potential design inputs.

Examine: A step in the ADEPT framework when the facilitator summarizes the team's findings for understanding and clarity.

Failure Mode and Effects Analysis (FMEA): An analysis tool for risk management that can be used to identify and prioritize potential failures, their effects, and their causes.

Features: The capabilities or characteristics of a product designed for customer use. Features are the tangible and measurable parts of a product.

Feature drivers: The reasons why the product offers certain capabilities or characteristics. They are the steppingstones to creating design inputs for the product itself.

Hazards: Top-down negative events beyond the scope of the Concept Space that can include biological, physical, and safety hazards.

Impact drivers: The factors that affect the impact of the feature or outcome on the user. Understanding these can help increase the value and positive experiences for the customer (or increase the likelihood or severity of the negative impact) and can lead to design inputs or service-related offerings.

Kano Model: A tool used to categorize and prioritize features based on customer satisfaction. It helps teams determine if features are mandatory ("must have"), One-Dimensional, Attractive, Neutral, or Negative.

L-shaped Matrix: Part of the matrix diagram used to list features and design inputs by row and column and evaluate their relationships.

Outcome Drivers: The underlying reasons or factors that can lead to the undesired outcome.

Parking Lot: A physical or virtual space used in team meetings to temporarily store ideas, questions, or topics that are not directly related to the current discussion. It helps manage the conversation flow and ensures these items are addressed later.

PCA Model: The PCA Model stands for Perception-Cognition-Action and is a framework to understand user interactions with a product. It suggests that every task a user does with a product requires perception, cognition, and action. This framework helps analyze use errors during testing and, in the context of concept development, to explore potential use errors to inform design inputs. It is used for both high-level functions and specific tasks during product development.

- **Perception** refers to what a user senses or fails to sense from the product's output, potentially leading to a use error. It involves the information users obtain from the design.

- **Cognition** is about how a user thinks and processes perceived information. Errors can happen due to gaps in knowledge or misunderstandings. Design inputs that focus on cognition aim to enhance understanding.

- **Action** is the physical response or control a user exerts. Errors happen if the user takes an incorrect manual action. Action-focused design inputs address the activity itself by simplifying or mistake-proofing it.

Problem Space: The area that focuses on understanding the problem before designing a solution.

Prioritize: A step in the ADEPT Team Framework that focuses on selecting and ranking ideas based on their impact on users.

Relationship Matrix: Part of the L-shaped matrix used to categorize the strength of relationship between features and design inputs.

Risk Priority Number (RPN): Within concept development, an RPN is a multiplication of severity and likelihood used to prioritize risks within a concept design. An RPN may be defined differently for other analysis.

Severity Rating Scale: A scale used to prioritize the impact of negative experiences.

Symptoms: Negative customer experiences that occur due to an unintended output or event. Symptoms are described as an outcome plus its impact.

System FMEA: A type of FMEA used to analyze the system as a whole.

Task Analysis: An evaluation of the steps a user must take to complete a task.

Tree Diagram: A visual model of a chain of events used to understand how one event leads to another.

Use Error: A deviation from what is expected when the product is used. This deviation can be from the designer's perspective or the user's perspective. A use error happens when a user does not perceive, understand, or act as expected, as in the Perception-Cognition-Action (PCA) model. It is use error, and not user error, so that we focus on how we design the product for use instead of blaming the user for making mistakes.

Use Process: The steps a customer takes to use a product. It includes the start and end points, as well as the intermediate steps and decisions.

Use Space: The environment in which a product is used, including when, where, how, and by whom.

Value-Added Analysis: A method to categorize process steps based on their added value. It identifies steps to prioritize, simplify, or eliminate.

Visual Model: A representation of a system, process, or concept using diagrams and charts.

Visual Template: A structured framework designed to facilitate idea generation and creative thinking within a team.

Voice of the Customer (VoC): The insights, expectations, preferences, and feedback that customers express about a product, service, or brand. Companies gather VoC data through interviews, social media, online reviews, surveys, and direct customer interactions.

INDEX

ABOUT THE AUTHOR

Dianna Deeney is an ASQ Certified Senior-Level Quality Professional with more than 25 years of experience spanning manufacturing and product design. She holds a BSME from Pennsylvania State University and is an IEEE Senior Member.

In 2017, Dianna founded Deeney Enterprises, LLC to bridge the gap between Quality tools and early-stage design. Despite their power, she saw these tools often confined to manufacturing and final inspection, limiting their impact on product development. While robust systems like Design for Six Sigma provide structured approaches, they may be impractical for many teams.

Dianna's mission is to offer practical, approachable methods that empower designers to collaborate, facilitate discovery, and translate insights into strong design inputs without disrupting established workflows. Her approach helps streamline development and reduce costs while ensuring products, from consumer goods to medical devices and industrial equipment, are inherently safe, dependable, and easy to use.

A recognized conference speaker and podcast host, Dianna champions "Quality during Design," promoting the integration of Quality tools at the earliest design stages. Through Deeney Enterprises, she provides educational materials, consulting services, online training courses, workshops, and coaching to help product designers enhance their processes.

Product development leaders and designers can learn more about Dianna's services and additional resources at DeeneyEnterprises.com.

Dianna resides in southeastern Pennsylvania, where she enjoys taking day trips to the mountains, shore, and city museums with her family and dog.

Let's PIERCE THE DESIGN FOG *Together*

For additional resources, including free learning materials and a list of services, visit **PierceTheDesignFog.com**

Interested in buying 10 or more copies? Contact us for our discount schedule: **orders@pingaugepublishing.com** or visit **PinGaugePublishing.com**.

Book Dianna Deeney for interviews or speaking at **support@deeneyenterprises.com** or visit **DeeneyEnterprises.com**.

Connect with DIANNA

🔗 diannadeeney

▶ @qualityduringdesign

✳ qualityduringdesign

THANK YOU *for reading*

If you enjoyed *Pierce the Design Fog*, please leave a review on Goodreads or on the retailer site where you purchased this book.

www.ingramcontent.com/pod-product-compliance
Lightning Source LLC
Chambersburg PA
CBHW071324210326
41597CB00015B/1339